P9-ASD-397

ANCIENT AGRICULTURE

ANCIENT AGRICULTURE

ROOTS AND APPLICATION OF SUSTAINABLE FARMING

Gabriel Alonso de Herrera

COMPILED BY
Juan Estevan Arellano

in association with the National Hispanic Cultural Center

TRANSLATED BY
Rosa López-Gastón

ILLUSTRATIONS BY
Bryan Romero

Ancient City Press

AN IMPRINT OF GIBBS SMITH, PUBLISHER

Salt Lake City | Charleston | Santa Fe | Santa Barbara

First edition 2006

10 09 08 07 06 5 4 3 2 1

Published by
Ancient City Press
An imprint of Gibbs Smith, Publisher
P.O. Box 667
Layton, Utah 84041

Orders: 1.800.748.5439
www.gibbs-smith.com

Designed by Blackeye Design
Printed and bound in Canada

LIBRARY AND ARCHIVES CANADA CATALOGUING IN PUBLICATION

Library of Congress Control Number: 2006926240

ISBN 978-1-4236-0120-3

This book is dedicated to the loving memory
of the greatest *jardineros* (gardeners) I've ever known—
my parents who raised me, Carlos and Lucía Arellano.
Que en paz descansen. (May they rest in peace.)

ACKNOWLEDGMENTS

THIS PROJECT WOULD NEVER have gotten off the ground had it not been for Orlando Romero, a historian and librarian who knows a gem when he sees one. I also owe a debt of gratitude to former state historian Michael Miller, who from day one recognized the importance of this project to the survival of Indo-hispano villages; Carlos Vasquez, director of Research and Literary Arts at the National Hispanic Cultural Center, for locating the perfect translator; Dr. Tomás Atencio of la plaza del Embudo, for the impetus to begin collecting the oral history and folklore of northern New Mexico; and Dr. Tomás Martinez Saldaña of Mexico City, for his expertise in agricultural history. I am grateful as well to Pilar López, of Cuenca, Spain, who, aware of my frustrated quest, at long last handed me a copy of Gabriel Alonso de Herrera's *Obra de Agricultura.*

To Richard Salazar and Mark Mondragón, I owe thanks for the initial translation. And now, ten years later, I wish to thank Gibbs Smith, publisher of Ancient City Press in Santa Fe, New Mexico, for believing in the project enough to get it moving like a grass fire, and also Marya Roddis of Los Ojos, eighty miles north, who introduced us. I was blessed to work with a wonderful editor, Ellen Kleiner, who, patient and tender like a master gardener, did an excellent job at making this difficult project blossom.

I also want to thank Dr. Carmen Trillo San José, of the Universidad de Granada, for sending me a copy of her book *Agua, Tierra y Hombres en al-Andalus;* Dr. Luis Pablo Martinez of Murcia, Spain, for his insights into regional acequias; and Dr. Thomas Glick at Boston University, a specialist in irrigation systems who in 1979 published a facsimile edition of Herrera's work in Valencia, Spain, which he graciously shared with me. *Gracias,* Albert Jaén, Tom Lozano, Dr. Gary Nabham, Habeeb Salloum, Joyce Guerin, and Liddie Martinez, and to Paula Garcia, of the New Mexico Acequia Association, special thanks for fighting against all odds to preserve the acequias.

Finally, this book could only have come to fruition with the help of my family: my wife Elena, who keeps me grounded in reality and supports me in all my endeavors, including sudden trips to Spain or Mexico, and burning desires to have her prepare exotic vegetables I sometimes find; my sons Javier and Carlos; and *mi corazón mi'ja* Única, who put up with me.

CONTENTS

INTRODUCTION TO THE ENGLISH EDITION

ANCIENT AGRICULTURE, the first English edition of Gabriel Alonso de Herrera's *Obra de Agricultura*, presents an approach to working with the soil that is as effective today as it was in 1513, when the book was first published in Alcalá de Henares, Spain. In our time of depleted soil, global warming, periodic droughts, mismanagement of water, and increased interest in organic farming, sustainable agriculture, and permaculture, Herrera's work both invites an examination of these practices from a historical perspective and offers timely information for implementing them. Herrera is known as the father of modern-day Spanish agriculture; less well known is the fact that his views can be seen as precursors of today's sustainable agriculture movement.

This English translation is based on the 1539 edition of Herrera's book, reissued in Spanish in 1998 by the Ministerio de Agricultura, Pesca y Alimentación (MAPA) of the Spanish government because it contained additional information the author had gleaned from the extensive reading and travels he undertook after 1513 and because it was the final edition he worked on before he died. Like a fine wine, his book improved with age.

His original book has an intriguing history, especially in regard to New Mexico's Spanish and Mexican heritage. Herrera, who was born in Talavera de la Reina, Spain, in the 1470s, was approached by Fray Francisco Ximénez de Cisneros, believed to be the chosen confessor of Queen Isabela in 1492 and in 1495 named Archbishop of Toledo, to write an agricultural instruction manual to help improve growing conditions in Spain, which at the time were rapidly deteriorating. Cisneros, an avid supporter of literature, dedicated his next twenty years to an ambitious publishing program designed to make select titles available to general readers, among them Herrera's finished product, *Obra de Agricultura,* printed by Arnao Guillén de Brocar. The original edition, which had graphics of vegetables and human figures, was dedicated to Cisneros.

The events of Herrera's life remain sketchy, but by 1492, according to scholars, he was in Granada, where he apparently remained for at least ten years. He had traveled to Granada with Fray Hernando of Talavera de la Reina, to whom he might have been related. It was in Granada that he received his extensive education in agriculture, a subject he had initially learned years before while working the land surrounding his boyhood home with his father. In Granada he increased his knowledge by observing Moors as they farmed and by reading works of Spanish Arabs that had been translated from the Arabic. Throughout this southeastern portion of Spain known as Andalucía, the Muslims had achieved agricultural advances that far exceeded any previously seen there.

The year 1503 represented an important turning point in Herrera's life, marking the start of his travels throughout Spain and much of Italy, France, and Germany, as well as the year he began writing his book on agriculture. Between 1503 and 1512, he traveled to Mantua and Rome, among other places, and in Spain to Vizcaya, the Pyrenees, Aragon, Málaga, Almería, Valencia, Andalucía, Granada, and Córdoba. During this time he also furthered his education by reading works by Greek, Roman, and Arab authors, becoming familiar with Aristotle, Theophrastus, Homer, Virgil, Hippocrates, Galeno, Pliny, Palladius, Columella, Séneca, Abencenif (Ibn Wafid), Avicena, Rasis, Mesne, and Crecentino. Returning in 1512 to Talavera de la Reina to finish writing his book enabled him to compare the knowledge he had learned in his village prior to 1492 with the information he had gained traveling and reading, culminating in the 1513 publication of his book. Thereafter, he continued to travel and read, as well as practice what he had learned, with the sole objective of improving the agriculture in Talavera.

Although little more is known about Herrera after the publication of the first edition of his book in 1513, some details about his life and the reception of his book can be determined indirectly through other known facts. For example, when he was between thirty-five and forty years old, his brother Hernando observed that many gentlemen had stopped reading their books on *caballería* (chivalry) to read his *Obra de Agricultura*, indicating the book's popularity, according to Spanish writer Isabel Moyano Andrés. Also, a note dated August 12, 1529, gives Herrera permission to publish another edition of his book, most recently printed the year before. And it is known that he was still alive in 1539, when another edition was printed, where he wrote, as he did in his previous one, "Recently corrected and added on . . . by the author himself." Though the year of his death has not been established, the most acceptable date is around 1540, since it had to have followed release of the 1539 edition and preceded that of the next edition in 1546, on which he apparently did not work. Between the first edition and the last one published while he was still alive, there were four other editions: one printed in 1520 in Toledo; another in Alcalá de Henares in 1524; a third in Logroño in 1528; and the fourth in 1529; the best edition of all, published in Madrid in 1818–1819 by the Sociedad Economica de Matritense de Amigos del País, appeared in four volumes. In addition, there have been eleven editions with the title *Agricultura General* and six editions translated into Italian.

Cisneros contacted Herrera to write such a significant book because of Herrera's broad perspective on farming, enriched further by his ability to combine the scientific agricultural practices of the early sixteenth century with his knowledge of old medieval treatises on the subject. It is thought that he was influenced, for example, by Ibn Wafid (1008–1074), the "Moro Abencenif" in his book, who wrote *Compendium of Agriculture*. Wafid, in turn, was a contemporary of Ibn Bassal, who wrote *Book of Agriculture* in 1075. Equally important, as a youth in Granada, Herrera had come in contact with numerous Moors who were excellent agriculturalists familiar with the farmer-writers of the School of Sevilla from the tenth to the fourteenth century, including Abu Zacaria Iahia Al Awan, whose classic *Book of Agriculture* of 1158 likely had an impact on Herrera, and Ibn Luyun, who in 1348 composed in verse *The Book of the Beginning of Beauty and the End of Knowledge Which Deals with the Fundamentals of the Art of Agriculture*. Among the agriculturalists that Herrera in turn influenced were Henri Louis Olivier in France and Conrad Heresbach in Germany.

According to scholars, Herrera wrote *Obra de Agricultura* in his birthplace of Talavera de la Reina, a village known for its tile work, probably with local farmers in mind. His father was Lope Alonso de Herrera, as stated in the book's prologue ("compiled by Gabriel Alonso de Herrera, son of Lope Alonso de Herrera"), a farmer "well educated and wise" about agriculture and apparently wealthy enough so that his three sons continued their studies. His mother was Juana Gonzales, who must have been deceased by 1528 since in the edition that year in Logroño, Herrera wrote: "And this I saw my lady and mother Juana Gonzales do, that she be in heaven." Scholars are of the opinion that she was baptized in the parish of San Salvador and buried in the parish of Santa María in the town of Talavera de la Reina.

In the words of Alvar Gómez, Cisneros's biographer, who wrote in the late sixteenth century, there were three Herrera brothers. The eldest, Hernando—born around 1460 and a professor in Alcalá de Henares from 1508 to 1510 and Salamanca from 1517 to 1518—was a respected man of letters. The middle brother, Diego Hernandez de Herrera, was a musician with a position as organist at San Ildefonso, the university church in Alcalá de Henares. Thus, Gabriel the youngest son, grew up with educated and talented siblings.

The three brothers were helped to attain their positions by Cisneros, who, in his quest to support culture, became their benefactor. In urging Gabriel to write his book, Cisneros was interested in publishing not so much a theoretical treatise but a practical work that would benefit the farmers immediately and simultaneously improve Spanish agriculture over time. Consequently, he not only paid for the publishing but also distributed free copies of the book. The urgency he felt was provoked by the expulsion of Moors from the peninsula and the arrival of the Christians, who appeared unable to produce enough food for the remaining inhabitants.

Scholars have differing views about how Herrera and Cisneros came into contact. It is possible that when Cisneros visited Granada in 1499, they met through Fray Hernando. Or Herrera might have been recommended to Cisneros by his older brother Hernando. In any case, according to various documents, Gabriel Herrera was in Granada in 1502 and 1503 at the service of the Marqués de Mondéjar, working on several *huertos*, or large vegetable gardens, and by that time already known to be more of an expert about agriculture than anyone of the era, including the Moors. Regardless of how they met, Cisneros became pivotal in the life and work of Herrera.

Although Herrera wrote his work at the beginning of the Siglo de Oro (Golden

Century) in Spanish literature, just decades before Miguel de Cervantes, Herrera was not concerned with a literary or scientific rendering but wanted to teach how to better work the land and produce more food, creating a manual for farmers about the agricultural realities of Castile and Talavera de la Reina. Thus, the book deals with the use of various tools and many agricultural practices, including amending the soil with compost, green manure, and animal manure, as well as the role of irrigation through man-made ditches, or acequias, which made Andalucía productive under the Moors. Still, because Herrera's work also includes information on the customs of rural life and superstitions pertaining to agricultural work, it serves as both a valuable source of information about agriculture and a bridge between historical and current agriculture. And since his prose is clear and elegant, reflecting the harmony of nature, it is accessible to farmers of both the past and the present.

Similarly, not only was Herrera's work very influential in Spain and elsewhere in Europe, there is evidence that is was known at an early date in other parts of the world, including the southwestern region of what is now the United States. According to personal communication with Dr. Joseph Sánchez of the Spanish Colonial Institute at the University of New Mexico, there is a 1777 edition at the Misión de Santa Bárbara in California that was published in Madrid under the title *Agricultura General,* with an inscription to Padre Fray Antonio Jaime dated 1797. Although no old copies have surfaced in New Mexico, the interviews I've conducted over the past thirty-five years suggest that his work was also known here. Certainly, many practices described in Herrera's book are identical to those still employed by native Indo-hispanos of the Río Arriba region—the area from the San Luis Valley in Southern Colorado to La Bajada, New Mexico, south of Santa Fe. For example, Herrera recommends saving seeds from the first harvest instead of the last and plowing at least three times before planting, the same advice given by the older farmers of New Mexico.

Surprisingly, although Herrera's book was written nearly five hundred years ago, most of the knowledge in it survived in the oral tradition of the Río Arriba region as it was passed down through the generations. More than likely, copies of the book itself were also transported up El Camino Real, the principal trade route from Mexico City to the northern part of New Mexico, by the friars who traveled with the first settlers, beginning with Don Juan de Oñate in 1598.

I first discovered Herrera's book when a friend, Orlando Romero, who was then head of the History Library at the Palace of the Governors in Santa Fe, called excitedly to tell me he had found an amazing book, which turned out to be a facsimile of the Spanish edition of Herrera's *Obra de Agricultura,* at Western New Mexico University, the only copy in New Mexico. I was astonished that the words of my father and regional friends jumped off the pages, prompting me to wonder whether elders of the region had been familiar with Herrera's work in the past.

Initially I was drawn to the volume because, like its author, I too had learned about farming from *my* parents, Carlos and Lucía Arellano, who passed on their extensive knowledge about plants and irrigation. Consequently, although I grew up in a family without much money, our place was a paradise full of glorious vegetables and fruits, including chiles, corn, cherries, pears, peaches, apples, watermelons, and cantaloupes. Then later, while a student at New Mexico State University and a VISTA volunteer in northern New Mexico, I became involved in oral history with La Academia de la Nueva Raza, gaining awareness of the vast knowledge about agriculture possessed by Indo-hispanos. In addition, I was an apprentice of a village *pícaro* (rogue), from whom I learned a great deal more about agriculture, especially ways to save seeds. I gained firsthand knowledge about land and water use and conservation from *mayordomos*, or caretakers of the acequia systems, not only in my own community of Embudo, which dates to 1725, but also throughout northern New Mexico as I traveled around the state. In a way, I was following in the footsteps of Herrera without knowing anything about him.

My relationship with the land became even more direct and personal after 1978, when my father passed away and my mother was getting up in age, making me aware that the torch had been passed on to me to continue working our family land. It was then, as I planted crops I remembered growing there in the past, that I realized the land had a memory. Although the agricultural extension service advised me that raspberries wouldn't grow in Embudo, I planted some regardless because I recalled them growing there before; and when my daughter was six months old, we harvested the first raspberries. Terraces my grandfather constructed long ago needed to be reclaimed, but first I had to learn how to do this from the late Cleofes Virgil, although I could relate to carving out earth because at this time I was creating environmental art and had already made a pond that when full looks like a lady in the water. While playing the mandolin, Cleofes advised: "It's very easy, just . . . throw one shovelful of dirt up and another down and by the time you get to the end you'll have your *ancon* [terrace]." I have since terraced all my property, and in one section have twelve terraces, which has not only simplified my irrigation but prevented erosion. I have grown as many as one hundred twenty-five different varieties of vegetables and have about twenty-eight varieties of heirloom apple trees, ten kinds of pear trees, and about forty peach trees. In addition, I grow *azufaifas* (jujubees), which are mentioned in Herrera's book; *níspero* (medlar); and even a persimmon tree, which while uncommon in northern New Mexico bears a very flavorful fruit here.

A huge apricot tree in what was my grandfather's garden produces a sweet nut that can be substituted for almonds, in addition to producing wonderful fruit. Like others who grew up here, my grandfather was a man of the soil. And from Pablo Romero, a longtime resident of Embudo, I learned surprising things about landscape design—for example, that grape plants, carefully arranged, can be used as *cortinas,* or curtains, in creating garden rooms outdoors. In 2001, our garden and my wife and eldest son's recipes were featured in *Saveur* Magazine, along with a photograph of my daughter Única on the front cover.

Eventually, Herrera's book seemed so relevant to aspects of agricultural practices today that I wondered about the possibility of translating it for modern readers. Then in 1997, while director of the Oñate Center in Alcalde, I was invited to give a presentation on Chicano literature at the Casa de las Americas in Madrid, Spain, which provided a necessary opportunity to obtain a genuine copy of this rare book. However, this proved to be more difficult than anticipated. Only after visiting every bookstore that friends and I could locate did we discover that a new edition was due to be published soon—which ultimately was sent to me by Pilar López, a friend from Cuenca, Spain.

After obtaining a copy of Herrera's book and studying it in some depth, my decision to germinate an English translation of the book was sparked primarily by my wish to nurture the roots of what I call the *agricultura mixta tradicional mestiza* heritage of agriculture in northern New Mexico, with which I had been familiar since farming with my father as a child. I felt that appreciating the roots of these traditions in northern New Mexico could help preserve our heritage. Additionally, I believed that the practices and direct style of Herrera's book would resonate with and be very helpful to present-day farmers of the region.

The next step was to find money to do the translation, as well as good translators familiar with the Spanish of Cervantes's time, which is similar to that still spoken in the Río Arriba area. Michael Miller, a former state historian, farmer, and writer from Cuarteles in the Española Valley, who worked for the organization that became known as the National Hispanic Cultural Center in Albuquerque, offered his help. I wrote a proposal to the Cuarto Centenario Committee sponsoring the four hundredth anniversary in 1998 of the arrival of Don Juan de Oñate in the Española Valley, and the project was funded.

In writing the proposal for the project, I had to first explore the ways a book written in Spain in 1513 could be presented as relevant for today's farmers and students of agriculture concerned with studying permaculture, sustainable agriculture, or modern-day agronomy or horticulture in a scientific world bent on globalization and genetic engineering in which 60 percent of our food supply is imported. First, I felt that Herrera's book could provide practical information necessary for increased food production in many areas. For example, very little of our food supply is grown in the Río Arriba bioregion. If a food shed is overlaid on the watershed, such as the whole Río Grande Basin, which in the United States includes Colorado, New Mexico, and Texas and in Mexico, Chihuahua, Monterrey, Coahuila, and Tamaulipas, it immediately becomes evident how little food is produced for local consumption.

Second, I became convinced that Herrera's book was more relevant than previously thought. It seemed to be of particular value to the drought-ridden Río Arriba bioregion of northern New Mexico, the heart of the Indo-hispano country initially settled by the Pueblo Indians and again in 1598 by Spanish colonizers led by Don Juan de Oñate, along with Tlaxcala Indians from the interior of Mexico.

The type of agriculture that has since developed in this region I call *agricultura mixta tradicional mestiza*, although its techniques, philosophies, and crops also are the result of global information flow. Even though many people erroneously refer to the type of agriculture in the region as "Spanish" in origin (meaning Castilian, Christian, European), this is not true. The land was actually worked not by the "old Christians" but by the Moors, who made the Iberian Peninsula produce exotic fruits and vegetables like never before, as documented by Dr. Zohor Idrisi in his essay "The Muslim Agricultural Revolution and its Influence on Europe" and other scholars.

To fully appreciate how the content of *Ancient Agriculture* relates to the local agriculture of northern New Mexico, it is necessary to first understand some aspects of the area's land and water use. One fundamental element of Indo-hispano agriculture is land grants, which are divided into commons, acequias, and *suertes* (lands irrigated by acequias). Other aspects are *huertas* (fruit and vegetable fields), *jardines* (gardens, reflecting the Persian concept of the garden of paradise), and *milpas* (cornfields), where Greco-Roman, Moorish, and Mesoamerican traditions all combine to form the *agricultura mixta tradicional mestiza* of northern New Mexico.

It can be termed *mixta* because it combines fruit trees, vegetables, and legumes, along with livestock, fowl, and bees. The mixed agriculture that developed in New Mexico around the land grants created by Spain and later Mexico, included both irrigated and dry farming, as well as grazing. Here, the *latifundias* (big expanses of land) of the Iberian Peninsula were replaced by the commons, and the peninsula's *minifundias* (small land holdings of only a few acres) by the *suertes*.

The agriculture that developed in northern New Mexico can be seen as *tradicional* because it has sustained Indo-hispanos for centuries and adheres to old methods of permaculture. Here acequias and terraces anchor the permanent agriculture of the system. This type of agriculture is also common to most places where the Spanish Crown established settlements, although it varies in different locations depending on the climate and soil.

The agriculture of northern New Mexico can be called *mestiza* because it combines Old World (Roman, Moorish-Middle Eastern) and New World (Mexican, Mesoamerican) systems (acequias and *chinampas*, known as floating gardens), techniques *(surco* and *tapanco),* fruits (cherries and *capulín*), vegetables (lettuce and tomatoes), and animals (chickens and turkeys). This mixed heritage is reflected in the agricultural terms of northern New Mexico. For instance, *huerta* is a Latin concept; *milpa* is a Mesoamerican idea; acequia is Moorish; *almacigo* (a seedbed later transplanted outside), common before the advent of greenhouses, is Arab in origin. A good example of how our everyday language reflects our mixed heritage is the following sentence, something a father might tell his son: "*Agarra la pala y haz un tapanco en la cequiecita*" ("Get the shovel and divert the water in the small ditch with a heap of dirt"). *Pala* (shovel) is a Latin word whose roots are Hebrew; *tapanco* derives from *Náhuatl tlapantli* (pile of dirt); and *cequiecita* (small canal to transport water) is an Arab word.

Although when researching the land patterns in New Mexico, it is customary to see their origin in *Recopilación de las leyes de los reinos de las indias*, or Laws of the Indies, of 1681, which are based on the *Ordenanzas* of King Philip II of 1573, we can, in fact, trace an Arab influence in all aspects of New Mexico land and water use, albeit under the guise of Roman law. When the Moors were expelled from Spain, their method of managing the land did not disappear but resurfaced in the laws under which the Spanish land grants were made to settlers of New Mexico.

Under the Laws of the Indies, the land was divided into commons (lands to be used in common by all the people affected by a particular land grant), the *suertes*, and the acequias (man-made ditches that deliver water to all the land below them). If the *suertes* are the body, the acequias are the veins that give life to the dry landscape. Above the acequias is the dry land, reflecting how land was managed in Northern Europe prior to the arrival of the Arabs in the Iberian Peninsula in 711.

Land grants usually applied to more than one *aldea*, or hamlet. For instance, in the Embudo land grant there are eleven separate hamlets, and all the historic names refer to the landscape: Cañoncito, Montecito, Apodaca, Bosque, Junta, Ciénaga, Nasa, Rincón, Bolsa, Rinconada, and the Plaza del Embudo (today known as Dixon).

Also, there are four main divisions within the commons, although they sometimes overlap: *sierras*, *montes*, *dehesas*, and *solares*. In Spain, *sierra* meant a mountainous terrain whose features resemble the teeth of a saw, but can also be from the Arabic, referring to a rugged high desert. It's in the *sierra* where the *cuencas*, watersheds, are located, providing water for not only the acequias but also the *norias* (from the Arabic word *na'ura*), or wells, for domestic uses. *Sierras* are also areas for gathering firewood and materials for constructing buildings. *Montes,* on the other hand, are harsh but habitable highlands.

The *dehesas*, known also as *tierras de pasteo*, are pastureland. And the *solares* are 138 x 138 foot sites where houses are often built between the acequias and the commons, away from the agricultural land. This is where settlers built their *corrals, gallineros, trochiles*, and *leñas* (corrals, chicken coops, pigpens, and woodpiles). Houses, if away from the town plaza, were constructed in an L- or U-shape, as were Moorish houses, and included a *dispensa* (utility room) and a *soterrano* (root cellar) where people kept food supplies for winter.

Irrigation of lands has always been a primary concern for arid northern New Mexico. In the Indo-hispano world, water is considered a *don divino* (divine right), which means it is nobody's property but belongs to all and ought to be divided equally among those who need it. Water is always divided according to the amount of land, based on the number of *peones* (workers) each landholder uses and the amount of water in the river, a practice prevalent also in other arid places worldwide. Water is distributed to all landholders by means of acequias, which are said to carry the lifeblood of northern New Mexico villages. To appreciate how acequias operate, it is helpful to compare them to the way blood flows in the body from the main arteries, to capillaries, to the blood vessels. Because settlers were aware of making the best of their environment, the *tomas*, where the acequias originate, were chosen for their *venitas de agua* (veins of water) that channeled water from the river. The *presas*, or diversion dams, were constructed to divert water from the river to the *acequia madre* (mother ditch). In northern New Mexico, the secondary acequias, known as *linderos,* are also used to delineate property boundaries.

Today, irrigation procedures in northern New Mexico remain much the same as when they were first established by the land grants. In spring when it is time to irrigate the *suertes*, each *parciante* (water rights owner) uses a *regardera* or *compuerta* (head gate) to divert the water from the *acequia madre* to their individual property. Once the water enters the *parciante's* property, it is spread out via *brazos*, which take the water to the different *ancones* (terraces), then further dispersed to irrigate the *eras* (beds known as waffle gardens among the Zuni people and Mesoamericans), through smaller channels called *ramos*. Eventually all the water comes together at the *desagüe* (outlet), which moves the water to the next *parciante*. At the last property, another *desagüe* sends the water not used back to the river, or sometimes to another acequia that uses the excess water, called *sobrante.*

When water is first run in the acequia in the spring, after the *limpia* (annual cleanup) is complete, the water that courses in front, picking up all the debris, is known as *puntera*. In the past this spring event was a festive occasion in villages, for it meant that the lifeblood of the community was flowing once again. Children would run ahead of the water yelling, "The water is coming!"

The person in charge of the water in the Río Arriba bioregion is known as the *mayordomo*. According to state law, the *mayordomo* must be elected along with the *comisión* (commission) and is under the direction of the commission. In earlier times he was known as the *cequiero* (the one who divides the water), for he acts like a "barmaid," making sure everyone has water. Further, the *mayordomo* is always referred to as one who is *"digno de confianza"* (worthy of being trusted), or *"el fiel del agua"* (faithful with the water).

Distribution of water follows the concept of *repartimiento de agua* (equal distribution of water), which is based on the Moorish notion of *equidad* that comes from the *Qur'an*, referring to the fact that regardless of the amount of water in the river, water is equally dispersed according to the amount of land under cultivation. Essentially, the water in each acequia is divided by the number of acres requiring irrigation. In drought years, however, people receive a paper note telling them when they can have the water and for how long, to ensure conservation of this precious resource.

Hilo de agua (thread of water), known colloquially as *filo de agua,* is an Islamic concept of water conservation. During times of drought, farmers are encouraged to water their gardens only when the plants *piden agua* (ask for water), which is when the leaves start to wilt, and to water early in the morning or in late afternoon but never during the heat of the day, a practice also recommended by Herrera. The only time a garden should be watered in the heat of the day is when a *parciante* is given a *papelito* (a paper note specifying when and how long each irrigator may water).

Two other important concepts associated with sharing water are *sobrante* (excess water), often used to irrigate new land for purposes of cultivation, and *auxilio* (a means for helping those who lack water). *Auxilio* occurs when a *comisión* from an acequia currently without water petitions for assistance from a *comisión* from an acequia with water.

The soul of the *agricultura mixta tradicional mestiza* of northern New Mexico is the irrigated land on which crops and gardens grow. It is divided into *suertes*, or long-lots that usually reach from the acequia to the river. The purpose of *suertes* is to give access to the river and to the commons, and thus to good land for growing crops and for grazing domestic animals. *Suertes* are further subdivided into *altitos,* or highlands where fruit trees are planted, and *joyas*, where fruits and vegetables are grown. Historically, *suertes* were allotted based on a lottery, or luck, starting from the center of a plaza. The idea of long-lots originated in the land patterns of Spain. In New Mexico, land division was oriented not toward growing for a market but rather to provide for the community, usually a close society based on family.

Terracing has always been an important concept in traditional New Mexican agriculture, a tradition ultimately derived from the terraces of the Alpujarras south of Granada and of Machu Pichu in Peru. Terraces serve a conservation function to retain the best soil for use in growing food, and also to bring water from higher to lower elevations in order to use the water repeatedly without the negative effects of erosion. Terraces all have their own names and characteristics, with the *altito* being the highest, followed by the *joya*, the *vega*, and finally the *ciénaga* along the riverbank.

Also, there are different types of terraces, those on slopes, in valleys, and along the river, where they are known as *ancones*.

Among the terraced, irrigated land, the *joyas* are the most fertile, and are where people usually plant their vegetables and fruits for home use or to trade. The *huertas* (large vegetable gardens) and *jardines* (small gardens) are both planted in these strips. No one would ever think of building on the *joya*, the most fertile land, which is set aside for food crops.

A *joya* can be further subdivided into *melgas* (smaller parcels of land between furrows) and *eras*. These sunken beds, as opposed to the raised beds familiar today, retain maximum moisture in desert environments, thereby conserving water. The threshing ground for wheat is also called an *era*.

Abono (compost) is another important concept in today's northern New Mexico agriculture and also discussed by Herrera, as well as organic growers and farmers who practice sustainable agriculture. Oral history interviews I have conducted further emphasize the significance of allowing weeds to decompose by putting them in a hole and then using the decomposed materials for building up the soil.

Considering all these aspects of the *agricultura mixta tradicional mestiza* of northern New Mexico, it becomes clear how Herrera's book can offer important information for today's farmer seeking to raise abundant crops despite scarcity of water or to practice sustainable agriculture or organic farming. The work can also illuminate the roots of New Mexico agricultural traditions, provide a sense of continuity with the history of other cultures, and encourage preservation of a long agricultural heritage. Its application, however, is not limited to New Mexico but rather extends to Southern California, Colorado, Utah, Nevada, Arizona, Idaho, Texas, Louisana, Florida, nontropical Mexico, Latin America, the Philippines, the Middle East, India, and other arid regions around the world where crops are irrigated by means of acequias.

This English edition, *Ancient Agriculture*, was originally translated by Richard Salazar, who has translated numerous documents stored in Spanish and Mexican archives while working at the New Mexico State Archives and Records Center in Santa Fe; Mark Mondragon, who has done extensive translating for the University of New Mexico; and myself. Richard Salazar translated the first third of the current book and Mark Mondragon translated the rest, while I pieced together the segments as integrally as possible. Later, to give unity to the work and make it more understandable to nonacademic readers, Rosa Lopéz-Gastón, a certified translator and interpreter, performed a final translation.

The challenge of how to present Herrera's work to today's sophisticated readers has been a central focus of all those involved with this translation project. At issue was whether to make a literal translation that would convey its original wording, resulting in language that would occasionally be awkward and difficult to read, or to make it more appealing to a contemporary audience. In general, we wanted it to be a work both valued by scholars of agriculture and also accessible to the general public, especially since water has become such a scarce and critical element in New Mexico's landscapes. We wanted the book also to be used by people who work the soil, the population for whom it was originally written. Keeping this aim in mind, while certain sections have been slightly abridged to make them more readable for modern audiences, we have made every effort to remain faithful to Herrera's original book.

Obra de Agricultura was divided into six books, whereas the English edition will be published in two volumes. This first volume contains the prologue and books 1, 2, and 6 of the original, which in the current edition are labeled chapters 1, 2, and 3. Chapter 1 focuses on ways to prepare the soil for raising vegetables and grains. It describes the types of land best suited to cultivating specific crops and when and how they should be planted, weeded, and harvested.

Chapter 2 deals with the soils, climates, and locations most favorable for vineyards, specifying which varieties of grapes lend themselves to different types of land and which locations are best for the various grapevines. It also addresses when and how to plant, weed, graft, and prune vineyards; explains the attributes of wine and how to make it; and provides instructions for building a wine cellar.

Chapter 3 presents the most advantageous times to perform agricultural tasks on a month-by-month arrangement based on phases of the moon. In addition, it establishes weather change indicators and examines other important requirements for optimal plant growth.

In conclusion, my hope is that farmers, especially those interested in traditional agriculture, organic farming, sustainable agriculture, and permaculture, will be able to apply Herrera's knowledge to working the land in ways that are productive, personally enriching, and ecologically valuable.

Juan Estevan Arellano
En el Embudo de Picuris, New Mexico

PROLOGUE

Opening statements to Obra de Agricultura

compiled by Gabriel Alonso de Herrera, son of Lope Alonzo de Herrera, and dedicated to the illustrious and magnificent Honorable Fray Francisco Ximenez Cisneros, Cardinal of Spain, Archbishop of Toledo, His Excellency

PHILOSOPHERS HAVE NOTED the difficulty of being at the forefront of anything from its inception. Building upon what has already been established, they note, is less laborious and hence not worthy of the same degree of reward or honor. Thus, with good reason inventors of any art are commonly referred to as the "fathers" of those who later practice it. Just as fathers are the source of their progeny in natural procreation, the originators are the source for those who succeed them in the arts, and therefore deserve the venerable name of "fathers," as substantiated by some very important examples.

Currently, we see first-rate masters in silversmithing and other types of metalworking who produce pieces from rough steel so delicate and beautiful that they appear to have been forged from pure silver, revealing finer workmanship than in the days of Tubal Cain, inventor of the forge. Similarly, tents are now more elegant and refined than those constructed in ancient times by Iabel. Nevertheless, sacred scripture refers to these men as "fathers" of their particular arts, not because others have failed to achieve greater perfection but because they were the first to practice them.

Inventors who originate a valuable endeavor should be held in greater esteem than those who later build upon it, given that such creation is both the foundation and key to the endeavor, Aristotle says; he who initiates such an endeavor is therefore worthy of great distinction and honor. Even so, I do not claim to be the primary inventor of the art of agriculture, which we can aptly call our "mother," for *she* must be viewed as the source of nourishment for our ancestors, ourselves, and future generations. Nonetheless, whereas exceptional books have been written about agriculture in Greek, Latin, and other languages, I can honestly say that I am the first to undertake the codification of the rules and art of agriculture in Spanish, although my effort may be modest, considering all there is to be said on the subject. It is a difficult task to reconcile the views of authors who at times disagree, all the while selecting some ancient and modern conventions while excluding others. There is no lack of scholars who state that the rules and advice of agriculturalists who wrote in Italy or Greece are not applicable to Spain due to the variations in soil and terrain, climate and conditions, or the effect of the stars. Such opinions, however, can be proven erroneous not only by reason and experience but by the expertise of the masters themselves, whom I intend to emulate. In fact, the Romans, and even Columella, who was Spanish, often adhered to precepts and views set forth by the Greeks and Carthaginians. Further, based on the preceding opinions, Spanish physicians practicing in other regions of the world should not be treating patients according to rules established by Greeks such as Hippocrates and Galen. Yet Italy, Germany, France, and other foreign nations benefit from the care provided by Spanish physicians. If the rules of medicine from different regions are generally beneficial for human organisms, whose conditions change on a daily basis, why are the rules set forth by agriculturalists with respect to working the land, which never changes, not the same?

According to Columella, and as is clearly manifest, God made the earth eternally fertile and infused her with the vigor of perpetual youth. Her ability to consistently bear fruit is inexhaustible, and age fails to deplete her strength—qualities that contrast with those of people and animals, with their clearly delineated stages of childhood, youth, and old age. Whereas elderly people cannot turn back the clock, despite the passing of time the earth, due to the benefit of lying fallow, remains useful and of service in a variety of ways.

Nor am I deterred by the views of those who maintain that any rustic farmer knows more about agriculture than Cato, Columella, Pliny, Palladius, and other ancient and modern writers, and even more than the erudite Marcus Terentius Varro or the celebrated Saint Augustine, who is revered above all Romans. In fact, Romans knew how to work the land so well (even better than Spaniards) because of their exceptional appreciation of it. Kings and captains alike cultivated the land with their own hands and were as honored for that endeavor as for their victories on the battlefield. The earth, in turn, flourished and produced much fruit due to the labor of farmers using plowshares worthy of laurels. Clearly, the individuals deemed to be most successful were those who valued their own labor.

Now, however, the land is cultivated by hired hands whose sole concern is their day's wage, or by careless servants, or slaves who despise their masters. Cultivating her carelessly is one thing, but it is quite another not to properly honor her, given that she is our mother. As a result of our carelessness, she seems to deny us a good portion of our subsistence. This is contrary to how the Romans, who were excellent farmers, related to the land. The agriculture books authored by Magon were translated from a foreign language into Latin by order of the illustrious Roman Senate so people could learn from his advice and information. This was deemed useful and necessary, a point on which Columella and I concur. Nevertheless, some people would maintain that the precepts and rules of agriculture cannot be codified and that there is no benefit to writing about them because the farmers, who stand to gain the most from this work, do not know how to read.

How ignorant such individuals are! They do not understand that they blame books erroneously. Their thoughtless view does not dissuade me, however; nor do I wish to respond to them, except to quote Pliny, who stated that no book is so bad that it benefits no one and that, at the very least, books occupy an idle person's time, preventing their involvement in vices that invariably result in scandals and sins.

In addition to the fact that Your Lordship—whom I always seek to serve and obey to the best of my ability—delegated the task to me, the principal impetus for my writing this book is that, God willing, it will benefit the public. There is no other income-producing endeavor more useful and beneficial than agriculture, not to mention its being inoffensive to God. By contrast, no occupation is more dangerous for the development of bodies and souls than that of merchants. Additionally agriculture, in its tranquility, security, and innocence, is superior to military life and has advantages over most other occupations as well.

Agricultural cultivation offers a secure and sacred life devoid of sin. This is so because rural life has many benefits: in the country there are neither grudges nor animosities, and health is safeguarded, thereby lengthening life. Although lazy men contend that rural life requires arduous work and reject it, in fact this is its most positive quality since working to sustain oneself serves God and promotes innumerable virtues of its own. We were born to work in this world so that we might rest in the next. Indeed, God does not want us to be idle but rather to work; he promises abundance to those who sustain themselves by the labor of their own hands, but not to those who eat due to the sweat of others. The farmer is thus exalted by God, and serving God, he reaps abundance for his family, whereas the indolent perish from hunger, paying for their sins. (I believe there is widespread hunger in Castile due to the laziness of the Castilian people, for many consume and destroy, while few work.) Work is considered so sacred that wise men instruct us to follow the example of ants that labor tirelessly for their sustenance rather than cicadas that waste their time singing and later die of hunger. Moreover, in addition to the fruits of the harvest the land offers great delight to those who reflect on her beauty, such as philosophers and scholars, who contemplate the secrets of everything.

Regarding physical health, it is well-known that before there were cities there were fewer ailments. And while there were not as many medicines available in the past, it was unnecessary to bring them from the Indies or Arabia because farmers cured afflictions with local herbs. Further, those who cultivate the land have hearty appetites; food tastes good to them and never makes them ill. In short, the earth gives us everything we need and is a vital necessity for sustaining life.

Agricultural cultivation is an inherently superior occupation because it results in three qualities not found in most others: benefit, pleasure, and honor. Just as living in the country ensures a healthy body, it also produces a good soul, while laziness leads to an unhealthy body, a damaged soul, and even an unpleasant disposition. Consequently, farming is the only occupation priests were allowed to practice under the sacred canons. We, like the priests, inherited this occupation from Adam, to whom it was ordained by God; are naturally predisposed to it; and depend on it for our existence. According to sacred scripture God created the earth, which in ancient times was highly valued and honored. As an expression of such honor, when the Romans prized someone highly they would assert with admiration that he was a good farmer. According to Cato, farmers were chosen to be soldiers and captains; as such, they preserved their innocence, lived virtuously, and succeeded in conquering most of the world. Some farmers were so esteemed they were promoted directly to the rank of captain, as Titus Livy reports regarding Cincinatus and as Emperor Trajan states in his writings. We too owe much to farmers, for their labor sustains our lives.

Gamblers, by contrast, endanger the very existence of the Republic—especially gamblers who call themselves noblemen. The public sins they commit through acts of gambling are highly contagious and damaging to those who emulate them because of their position in society. To prevent further corruption, we must recognize that true nobility resides in the soul and is given by God, not derived through circumstances of birth or position in society. Thus, true nobility becomes manifest in virtuous actions, not arrogant words. Those who claim nobility because of ancestors and then commit vile deeds, greatly undermine their status, compromised already by the fact that they never achieved heroic feats. Actually, individuals toughened by working in the fields are more capable of heroic feats, since they have been raised in the sun, rain, snow, and wind, performing tasks both day and night. Sharp swords cannot cut their flesh, for it is as if they were wearing protective gear.

By contrast, those who stay indoors, protected by shade like ladies of the court, have the constitution of women. They may claim to be fiercer than lions in cities, but if taken to work in the fields or perform military exercises, they can be quickly debilitated by a simple cold. Further, they require more equipment for their comfort than is necessary to lay siege to cities or fight their enemies. This is why in the art of war people from the country are preferred over those who lead idle aristocratic lives or do women's work, immersions that weaken both constitution and spirit.

So while it said that nobility and farming are mutually exclusive, they are actually one and the same. Not only were kings once farmers, but farming is a noble endeavor that is not the least bit servile. If it is scorned in our times, such disdain is due not to agriculture's inherent qualities but to Castilians' inability to read about its benefits.

Your Lordship selected me to write this book. In addition to being an admirer of the agricultural way of life, I wish to help people. Thus, I will dwell not on the difficulty of the enterprise but on its benefits—the spirit in which I enthusiastically accepted the task. I also am aware there may be others capable of presenting this art in a finer style and with greater wisdom. That said, I wish to conclude by noting a few rules for success in farmwork. Favorable outcomes in farming, as in any other noble endeavor, depends on knowledge, ability, and dedication. If one of these factors is lacking, it is better to have experience without art than art without experience. Positive gains also require an affinity for agricultural endeavors, meaning a farmer should value and honor them as tasks calling for diligence. Moreover, no individual should attempt to farm inferior land, because it is extremely costly and results in little benefit. It is far more preferable to use such lands for less costly, more beneficial endeavors. Farewell!

CHAPTER ONE

PREPARING THE SOIL

Beneficial lands for farming, cultivation, and crop productivity

LAND IS THE SOURCE OF ALL THINGS, and because of the great benefits derived from it we call it "mother." Based on its location and features, all land falls into one of three categories: plains, valleys, or mountains. Each mountain has two components: slopes and a peak. Valleys are richer than plains, and plains richer than slopes and peaks, because slopes and peaks constantly provide substance to all that lies below. Consequently, the bases of the hills are better suited to farming than the slopes or higher elevations, simply because they are richer lands.

Moreover, the soils themselves have specific properties that result in their being very good (productive), very bad (unproductive), or between the two extremes. The better the soil, the more often it is cultivated.

In addition, due to geography and climate some lands are very hot or very cold, while others are temperate, meaning between hot and cold, humid and dry, and between rich and sandy soils. It is not my intention to discuss the extremes that contribute to soils' productivity or lack of it. Rather, since incurable illnesses are found in lands just as in the human body, I will provide indications for determining which lands should or should not be used for various types of cultivation, as well as procedures for improving the productivity of soils to support the growth and harvesting of grains, legumes, spices, and other crops.

Methods for Identifying and Improving the Quality of Soil

All who have written about agriculture agree that farmers must know the quality of the soil they are farming so they can judge its requirements, the crops that can be successfully grown on it, and ways to improve it. Although some lands produce annually, not all soils have the capacity to achieve this without irrigation and fertilization. In many places, such as the kingdoms of Aragon and Valencia and the kingdom of Granada, if the soil is not softened by rain, farmers resort to artificially irrigating the wheat fields; in others, especially where the soil is very moist, they thinly fertilize the lands before planting a variety of crops. Everywhere, successful production hinges initially on the application of certain sowing, weeding, plowing or digging, harvesting, threshing, and storage techniques. First and foremost, however, the farmer's task is to study the land and its many characteristics.

SIGNS FOR IDENTIFYING BAD AND GOOD LANDS

Agriculturalists who recommend signs for identifying the fertility or sterility of lands agree that, in general, heat is not a sufficient indicator. In fact, many statements commonly made about assessing the quality of soil may not always be true. For instance, it is often said that darker soils are best for wheat. And while for the cultivation of this crop they are better than other soils, since they endure heat better and tend not to flood, color is not a guarantee of good lands. Crecentino, Columella, Palladius, and Pliny all state that good land suitable for wheat cultivation should be sticky, soft, and not sandy. Soil can be tested for these qualities in the following way. Take a small sample of soil and wet it with saliva or water, then roll it between your fingers. If it sticks like dough, it is good and rich, but if it is rough and sandy, it is not good soil. Of course, we are not referring to potter's clay, because although that is a sticky type of soil, it is not suitable for wheat since it is extremely hard and dry.

Another sign for identifying the quality of soil is how quickly it absorbs water and how long it retains the moisture: the more it retains, says Palladius, the better the soil. Good land becomes soft, spongy, and dark when it absorbs rain or irrigation water, rather than hard or whitish, which indicates that it is not very fertile. For a constructive experiment, dig a hole big enough to plant a tree, and replace the soil within two or three days. If there is more than is needed to fill the hole, the soil is rich and good; if there is just enough to fill the hole, it is of medium grade; and if there is not enough to fill the hole, the soil is thin and very light.

The quality of soil can also be assessed by studying the characteristics of water that flows from it. From good lands come sweet waters with good taste—even if the waters are murky. While most clear waters come from thin, even sterile, lands that are sandy

and rocky, murky waters are a sign of lands as rich as the lush grasses that grow in them, just as fatty meat makes a thick, rich broth. An exception occurs when murky waters come from gypseous lands, which are neither rich nor fertile; worse, the mire emanating from the gypsum has a stench.

When water does not flow from the land being assessed, it is possible to use another method for determining the soil's fertility, such as the following. Take chunks of soil and dissolve them in a container of sweet, good-tasting water. Let the soil settle, then pass it through a sieve, after which the water's appearance and taste will indicate the quality of the land. If the water is clear and good tasting, the soil is likely to be fertile, having acquired the attributes of the water. This is particularly true of lands used as vineyards.

Further, a good sign of the fertility of soil is if it is well-covered with grama grass and herbs, or with naturally occurring dwarf elders, rushes, bramble patches, broom thickets, clover, ammi visnaga, mountain sloe, hemlock, giant fennel plants, reeds, large thistles, hollyhock, and young oaks, and if all the herbs and plants that grow around them are succulent and vibrant shades of green. Gravelly lands, and those supporting thickets of rockroses and oak saplings are moderately fertile. Oak groves tend to grow on lands that are too sandy for other plants—so much so, they are characterized as having "dead sand." Likewise, places where rosemary and heath thrive are often too light and sterile for other plants, as are lands that are barren.

Lands free of grasses or herbs are incapable of producing wheat. Lands insufficiently fertile to sustain grass are even less capable of growing wheat. Similarly, fields that grow esparto grass are generally too dry

to support wheat, the waters emanating from them are salty, and very little can be done to improve them; the few possible remedies would be very costly, difficult to implement, and thus not worthwhile. Some of the esparto lands lacking substance and fertility, however, are moderately good for wheat—or if not wheat, then rye or other seeds that thrive in light soils. Despite the foregoing indications, it is not always possible to fully determine the nature of lands from plants, so to be more certain of making accurate assessments, it is necessary to consider all the signs and tests for good lands.

DEFICIENCIES OF THE LANDS

Remedies to correct land deficiencies can enhance agricultural productivity, but only if the deficiencies are limited. If lands (*suertes*) are fertile, with springs or other natural water sources, but are too moist to be suitable for wheat cultivation unless they are drained, the problem can be solved by creating drainage channels, or *sangraderas*, through which the water can be removed.

Just as bleeding or removing unhealthy fluids often rids the human body of offending substances, soil can be drained to eliminate harmful moisture. A diligent farmer will ensure that the drainage channels are appropriate for the type and quantity of moisture, neither drying the land more than necessary nor leaving undesirable moisture. Even large lakes can be drained using this method, as agriculturalists have indicated, and as I saw in the Mantua region of Italy, where lakes were drained using channels and ditches so wheat and herbs could be grown there.

There are two drainage methods. The best method to use when there is a large amount of water is construction of a series of small ditches that drain into an *acequia madre* (mother ditch), though sometimes a single large ditch is sufficient. The other method which, according to Palladius and Columella, is preferably used when there is less water, particularly no springs, is the digging of *ciegas* (hidden ditches). These can be dug as deep as necessary, then the bottoms lined with pebbles and round rocks that are subsequently covered with soil, making everything level so the ground can be planted yet the water drains.

If the soil is moist and cold, Crecentino adds, it is beneficial to mix in a large amount of argil, as this will produce heat. Pliny argues against this remedy, pointing out that the cost exceeds the gain. Since argil is sterile, and to grow wheat or fruit the sterility of soil must be corrected, in my opinion it is better to use plenty of fertilizer, which is by nature hot and thus easily able to remove moisture from the land. But where fertilizer is used—whether to assist in the cultivation of wheat, barley, fruit, herbs, or vineyards—the flavor and preservation of the produce will be impaired because fertilizer is a corrupting factor, particularly if it is used extensively in places other than gardens with abundant moisture.

Other deficiencies include soil that is too thin or too rich. Very thin soil does not have enough nutrients to sustain wheat, and in very rich soil wheat fails to grow, eventually turning to straw. If soil is only somewhat thin, the best remedy is to let it lie fallow for one year and produce in the next, alternating as necessary. For very thin soil, agriculturalists note, the best remedy is to let it lie fallow over an extended period and then produce grass with the aid of fertilizer, allowing livestock to feed on it; in addition, burning the entire area will make the soil more productive. And for very rich soil, some say, it is good to plow in enough sand to break up the soil, especially if it is sticky. In my opinion, however, this remedy is not much better than the use of argil. Soils that are very rich should be sown continuously, year after year, to lighten and improve them. In some years, provided the land is sufficiently level, it is good to sow plants that thrive in rich, tough soil, such as broad beans, melons, cucumbers, grains, millet, and similar plants, since these crops can survive almost regardless of the air and soil in which they are planted and all the while temper the soil so it will be primed for wheat.

If farmlands are very rocky, making the soil difficult to cultivate, gather the rocks and pile them out of the way or along the edge of the field. If there are only a few rocks, leave them in place so that during the summer they can protect seedlings from the sun and keep their roots cool. If farmlands are covered with rushes or grama grass, burn these plants, since otherwise they will continue to grow with the cultivation of the soil. As for ferns, Palladius states that planting lupines or broad beans among them can eradicate them, or they can be cut with a shovel when they are small. He also affirms that nothing destroys them more effectively than their own sap. Pliny explains that if for two consecutive years ferns are cut before they produce leaves, the roots will dry up. If they are cut with a cane-cutting knife, this should be done when they are seedlings. Further, it is said that dwarf elders and ditch reeds can be eradicated in the same manner.

Actually there is no better way to deal with the soil and climate deficiencies of some lands than to work *with* them, cultivating what is suitable, as Virgil affirms. In hot areas, cultivate plants that grow well in heat. In cold regions, sow plants that will not freeze. In dry places, plant those that do not require much moisture or care. In humid areas, sow plants that need more moisture. In higher elevations or mountains, grow those that require more air. In valleys or lowlands, sow varieties that need protection from the wind. In shaded areas, plant those that require little sun. In sunny places, grow those that do not thrive in the shade. Accordingly, lands not suited for wheat can be used for herbs, orchards, or

many other useful crops, each with its own distinctive attributes, all graciously given to us by God.

Unfortunately, the failure of farmers to use each type of soil in the most suitable manner and not leave many lands uncultivated explains why Spain is so poor. Because farmers could derive far greater utility from the land than they do currently, in the following sections I will elaborate on numerous topics that can potentially enhance their knowledge about types of crops and methods of cultivation to attain greater productivity. I will expound on different plowing techniques, indicating when they should be used, and on various methods and times for fertilizing lands utilized for a variety of crops. I will also discuss irrigation timing and techniques. In general, lands that are not irrigated and are temperate are those most likely to consistently bear fruit, since there is little that can damage them, while rich soils can be ruined by dryness and thin soils by too much water.

Benefits of Plowing or Digging

Proper plowing or digging is essential for productivity of the land. To be effective, these tasks must be done in accordance with appropriate timing, techniques, quality of the soil, and characteristics of the region. Moreover, some lands must be farmed in a certain manner, while others require different methods. Some also need more work than others.

Crecentino asserts that plowing or digging yields four main benefits. First, it works the land. Although he speaks of "covering the land," it is more accurate to think in terms of "opening the land," because plowing or digging the soil allows sun and rain to penetrate it, priming the land for cultivation. Second, plowing or digging levels the land. Some places are naturally higher than others or have more holes, and when it rains heavily plants in areas with holes can drown, or in times of drought those on higher ground can dry up. Plowing ensures that the soil is as level as possible, so water flows evenly when irrigated or in rainstorms, and the sun will warm it uniformly. Third, plowing incorporates heavy soil with light, and fertilizer and seeds with soil. Plowing integrates everything properly, especially smaller seeds like wheat and barley, which otherwise might remain uncovered then dry out in the sun, freeze, or be eaten by birds, and thus not sprout. Fourth, plowing breaks down large clods of earth so the soil can better sustain heat and moisture for greater productivity. Accordingly, agriculturalists advise that whenever fields are plowed, the soil should be completely broken down.

In addition to yielding these four benefits, plowing or digging results in two other fundamental improvements of the soil. First, it eradicates weeds, which if allowed to grow can sap the substance of other plants, choke them, and even kill them. Second, it modifies the soil in the event that, unworked for a while, it has become hard.

Every field that is to be sown should first be leveled according to its soil type. When farmers do not know how to plow or dig properly, however, or the right times in which to do it, their work can be in vain or, worse, detrimental to cultivation. Crecentino, Columella, and Palladius assert that soil can be seriously damaged if it is plowed or dug up when it is either very wet or very dry. In the former case, instead of becoming spongy and soft it becomes claylike, hard, and resistant to plant growth for several years. If sown in this state, much of the seed will be lost, and the seeds that do grow will take a long time to sprout and not grow well. If the soil is extremely dry, on the other hand, when plowed or dug up it is likely to leave large clods of earth on the field or garden. Further, if the soil is overly fertilized it cannot be well prepared because, even if it is worked when it is not very wet or very dry, it will require twice as much labor. Also, the plow or hoe should not be covered with a claylike soil, but instead a very fine, sandlike soil should fall easily off the tool.

All stubble on the land that was productive in the previous year should be burned to derive maximum benefits. Additionally, all land designated for wheat or herb cultivation should be burned to first clear it of brush, brambles, and twigs. Stubble from previous crops should be burned when winds will not carry away the cinders or ashes, which are extremely valuable to the soil and likely to cause damage if spread elsewhere. The same method should be used when clearing brush, brambles, or twigs, which should be done during cloudy or humid weather so the coming rain will mix the ashes with the soil, invigorating it. When burning land designated for wheat production, it is best to do so while the weeds are dry so their seeds are also burned; this way they will not grow again and choke the wheat. Despite all the benefits of burning the land cited by Virgil in his *Georgica,* he omitted an important one: if the roots of weeds and plants burn, the ashes will fertilize the soil.

The best time to plow is after the stubble and brush have been burned, and it has rained or the land is wet, especially if plowing lands that are heavy and humid. According to Theophrastus and Varro, it is most productive to plow during hot weather. This is particularly true for heavy soils, because the heat scorches weeds and their roots. Soils that are loose, sandy, thin, and lack substance should be plowed at least twice, one time in the spring, according to Cato, when pear trees are in bloom. Theophrastus maintains that loose, light soils should be plowed in winter, so that the hot sun of summer will not dry out the little substance they possess and so the soils will more likely absorb the moisture. If rich lands are not plowed after being harvested, it is best to wait until winter, at which time plowing will cause the weeds to surface so the coming frosts will kill them. Plowing later, once the weeds have seeded, would result in weeds that are planted instead of destroyed. It is useful to plow and replow to eliminate weeds, particularly if the soil is hard, because then it will be finely ground rather than loose and weedy, a view confirmed by Crecentino. Rich soils require at least three preparatory plowings and a fourth at the time of sowing, as Virgil corroborates when he writes, *"Quarto seri sulco"* ("Plowing four times").

The precise number of times soil should be plowed before sowing varies somewhat with location and type of soil. According to Pliny, in Tuscany lands are so hard and full of weeds that they are plowed nine times before being sown. When hard soils are plowed three times, the first plowing—called *"alzar"* or *"barbechar"* in Spanish—is done about a month before sowing so they can absorb needed moisture. The second plowing, done in the spring, is called *"binar"* ("a second time"), a term derived from the Latin word *bis* ("twice"). The third plowing is done shortly before sowing, to spread and cover the fertilizer if the soil and air require it—a procedure less necessary in hot, dry lands than in those that are cold, humid, and irrigated. During the third plowing, the plow should dig as deeply as possible to thoroughly mix the bottom soil with the top, making it more productive, and the soil that is dug up should be left so level that the plowing is almost imperceptible.

The fertilizer should be old, rotted, and spread only lightly to decrease or eliminate the weeds, and it should be covered so the sun does not damage its potency. If the fields are at a distance, making it difficult to haul fertilizer, livestock such as goats, cattle, or sheep should be driven there. If possible, fertilizer should be spread shortly before the soil is sown, not only to preserve its potency but also to prevent the growth of weeds. At the time of sowing, the soil should be soft, spongy, and finely ground rather than hard and tough, so wheat will grow with longer roots. Columella asserts that those who plow should be tall so they can better handle the plow, a principle Virgil calls *"aratorin curvus."* Further, the person who plows should carry a knife, so when the plow is lifted he can cut any roots that are caught in it, thus preventing harm to the pulling oxen.

Columella also states that the furrow should be neither too long nor too short. If it is too long, the work will be extremely tiring, and if too short much will be lost with the oxen's hoof prints when making the return. It is said that for heavy, hard soils the furrow should be 120 feet in

length; for loose, light soils it can be longer, while for soils already planted with crops it should be shorter. The farmer should always maintain an even depth in the furrow, and the plow should be pulled as straight out of the ground as possible. The farmer should additionally plow the furrow without interruption, because once oxen have been trained to plow an entire furrow, stopping confuses them. Therefore, it is best to travel the length of the furrow without pausing, giving the oxen a brief break before making the return. Allowing them to regain their strength through periodic breaks at the end of each furrow will encourage them to work each subsequent furrow faster and more efficiently.

According to Columella, soil that is hard is plowed better with oxen than with mules, because the former is stronger and easier to control. Oxen also work better and more efficiently than mules, particularly if the soil is loose, level, and without plants.

Regarding the best time to plow, the prudent farmer should be aware that in warm regions the effects of summer begin before they do in colder areas, just as in cold areas winter begins to take effect earlier than in warmer regions. Consequently in warm regions, the soil should first be plowed shortly after Christmas, and in colder areas, in March.

Moreover, most agriculturalists maintain that it is preferable to plow when winds are blowing from the northwest or south. Such warm and humid winds, Crecentino says, provide substance and season the land. According to Varro, as the wind picks up, usually from the beginning of February until the end of March, it prepares the soil. Even though these winds are felt throughout the year, they are never as constant or beneficial as at this time. It is not recommended to plow when there are cold northerly winds, Pliny affirms. These extremely cold and dry winds are harmful to farmlands, although they are useful for eradicating weeds if the land is plowed or dug up, exposing the roots of the weeds to the cold air and thus freezing them.

Having discussed the best times to plow, we shall now address techniques. Plowing along a hillside should be done laterally, because plowing from top to bottom would be extremely hard work for both the oxen and the individual; it is also harmful to the land, since furrows dug from top to bottom are incapable of absorbing the water as it rushes downhill after a hard rainstorm, taking the topsoil with it. If the majority of cultivation is wheat, one should make certain that the furrows are plowed closely together, as well. In this case, farmers say, ridges between furrows are allowed, but the soil should be examined closely to ensure that it is finely ground and nothing is hard. If one can put a *vara* [Spanish measure for 0.84 meters] through the soil lengthwise, it has been well plowed; if the measure becomes stuck and covered with soil, it should be replowed. Multiple plowings should always be done in the form of a cross—that is, the furrows of the second plowing should run opposite those of the first, and the third opposite those of the second.

With regard to oxen pulling plows, Columella claims that it is better to harness them by the neck than by the head or horns, so they use their entire bodies to pull, providing greater force. It astounds me that he believes this, given that he was a Spaniard and knew that Spanish oxen are large and have great strength in their heads, allowing them to easily pull a plow. This advice, it turns out, is more applicable to oxen in France or Italy, which are small and have thin horns. Pliny advises that oxen should be very tightly and firmly bound together when plowing, forcing them to carry their heads high, which would be impossible if harnessed by the neck.

Farmers who use collars for plowing animals, whether oxen or mules, must utilize headstalls, especially when plowing vineyards, orchards, or other planted areas, to prevent the animals from causing damage. Oxen that are harnessed by the horns cannot cause damage unless they stop. Also, collars should fit loosely around the neck, so as not to hurt this area. Columella adds that when animals are wearing collars, the plow should be pushed forward at the end of each furrow, because otherwise movement of the collar could cause injuries difficult to cure.

When unharnessing animals, they should not be released until their heads, faces, necks, and especially the places where they were bound together have been rubbed, a practice that is pleasant for the animals and also makes them more gentle. Columella recommends that when unharnessing animals, the skin which has been wedged into the flesh should be loosened, both on the head where the yoke was placed and on the neck where the collar was attached, to avoid serious sores, particularly on the neck. One must be as kind as possible to the oxen, because of all the animals we use they help us the most, work the hardest, and are the most valuable.

Finally, to cite Virgil's advice and that of most other agriculturalists, as we Castilians say, "*Quien mucho abarca poco aprieta*" ("He who undertakes too much, retains very little"). I agree—farmers should cultivate only what they can properly care for. One field or vineyard that is well tended will be more productive than three or four that are neglected.

Advice on Sowing and Weeding

Farmers, having dedicated much labor and cost to leveling the land, should be present when the sowing is done, as this is more important than any other work in the fields. Agriculturalists, Palladius in particular, advise that the owner's presence enhances all labor. Crecentino adds that the owner's presence heightens and improves everything that is done and that there is not a laborer in the field or a team of oxen as important as the presence of the owner or the person in charge. Magon suggests that anyone who buys or inherits a farm should sell their home in the city and go live on the farm. I maintain that this is of utmost importance when sowing and weeding, since otherwise it is all too possible to visit the fields and be confronted, too late, with failure due to sterile seed, seed damaged in silos, excessive rain, frost, an unexpected heat wave, seed eaten by birds, or pilfering.

THE ATTRIBUTES OF SEEDS

Farmers should always attempt to obtain the best possible seeds. From excellent seeds grow excellent fruits, while good fruit seldom grows from bad seed, unless it is assisted by extremely good weather. Crecentino, Virgil, Varro, Columella, and Pliny all say that, more than anything else, seeds for sowing should be fresh. Generally, those more than a year old are not good; those two years old are worse; those three years old are poor; and seeds older than that are sterile. While this is true for seeds of wheat, rye, barley, and some vegetables, other seeds improve substantially with age.

Palladius adds that seeds should be selected based on their proven success and productivity in the soil where they will be sown. Just as we look for good stock in animals such as horses, dogs, and bulls, so must we do the same with seeds. Regardless of type, seeds should be very full, heavy, unwrinkled, not moist, hard, free of other seeds or weeds, and if possible, individually selected.

For optimal cultivation, it is best to select good seeds from good lands. Seeds that come from a humid area, for instance, can easily spoil. But it is also true that poor soil and even changes in climate can transform good seeds. For example, if good wheat is sown in very cold, thin lands, within two years it will turn to rye. Preferably the seed should come from lands that are similar to those being sown: hot to hot, cold to cold, dry to dry, temperate to

temperate, humid to humid, and so forth. If it comes from very cold lands to very hot, from rich soil to thin, from very moist to very dry, or vice versa, the seed will not grow as well. In addition, the seed should not come from a sprout, because these are often sterile.

Wheat seeds should be selected from grains that are full and golden. Regarding white wheat, which has thin grains, it is ideal if they are somewhat golden and, when split, display the same coloration inside as outside; they should also have a good scent and not be mealy. Further, it is preferable that grains come from a well-ventilated granary rather than a cellar. If the grains come from a silo, they should be dry and not have been kept there so long as to have become grub ridden or eaten by worms. Columella adds that when wheat is cleaned, seeds that have fallen to the bottom of the pile are best, because they are heavier. A sieve, used for such selection, allows the light and infertile seeds to remain on top like straw, while the heavy ones fall to the bottom. It is also helpful to set aside grains with many spikes, as these indicate healthier seeds. They should be cleaned and their seeds planted separately from the others every year until there is enough of such seed available to plant an entire field.

Finally, all the seed sown in a single cultivated field should be of the same type. There are people who, fearful of lean years, combine many types of wheat seeds (white wheat, spring wheat, and other varieties) and sometimes rye, in the hope that something will grow, but even if everything grows well, it causes confusion and is not beneficial. Others attempt to ensure a crop of wheat by sowing different wheat seeds in alternating rows, thus separating each type in the field. I disagree with this method too, however, because each variety of wheat requires its own type of soil. Spring wheat grows better in lowlands and plains with rich soils; darker wheat prefers higher elevations; and white wheat favors light and humid lands. Hence, each variety should be sown separately in the soil that best suits it. Another reason seed should be of a single type is that different varieties sprout, grow, and ripen at different times.

Similarly, native plants should not be integrated with foreign ones, according to Theophrastus, who recommends that if there are no fresh seeds to sow, older seeds should be carefully examined to determine if they are viable, using the following technique. Place the seed in warm water, ensuring that it is not too hot, which would scorch it. If it grows, even if it is not a high-quality seed, it should produce an average-size plant. If it does not grow, however, send it to the mill and use it in the most productive way possible. Despite variations in where the seed is sown, it invariably grows the same plant, as Theophrastus asserts. Now that we have selected the seed, we shall, by the grace of God, begin to sow.

CONDITIONS AND TECHNIQUES FOR SOWING

Hesiod and Virgil, among the most important individuals who have written about agriculture, maintain that farmers should know a great deal about the influences and properties of the heavenly bodies, particularly with regard to the sowing of crops. It has been well established that their influence affects the earth, causing seasonal fluctuations, with effects varying from year to year. Thus, some years are humid and wet, others dry, and others moderate. In some years, seasons arrive early, in others late. Some years are unproductive and others very generative. Undoubtedly, farmers could acquire beneficial knowledge from astrology, permitting them to predict the characteristics of forthcoming seasons, but we cannot expect uneducated farmers to attain such knowledge when numerous scholars and medical doctors who should be aware of this information know almost nothing about it. In chapter 3, I will set forth some of the celestial signs indicating the characteristics of forthcoming seasons—such as humid, dry, rainy, or windy—according to the teachings of remarkably erudite scholars, while here we will look specifically at their effects on sowing.

Above all, farmers should be mindful that sowing, as well as grafting and plowing, must be done as much as possible when the moon is waxing, preferably at the beginning of the cycle. According to Crecentino, the moon's two quarters of waxing are for activities related to growth, while the two quarters of waning are for activities related to consumption. The first quarter characterized by heat and humidity, fosters the growth of seeds and plants far more than the second quarter, which is hot and dry. During the two waning quarters, it is cold and humid, and affects seeds more than trees. If trees are planted in the late afternoon when the moon is waning, they will grow very well because they will have the entire waxing period in which to mature in addition to receiving the transplanting benefits conferred by the waning moon. This principle, however, does not hold true for seeds. If seeds are sown during the waning moon, they generally do not sprout, and if they do grow they are neither as good nor as hearty as they would otherwise be. Pliny and Crecentino agree on the benefits of sowing during the waxing moon, and Palladius mentions it twice. Hesiod states that the ninth and thirteenth day of the moon are the best days to plant trees, but are not ideal for sowing seeds. Juan Pico, the Count of Mirándola and an eminent scholar in all sciences, also acknowledges this fact.

In addition to considering celestial signs, farmers must be aware of climate and soil conditions conducive to sowing. They should recognize if the lands they sow tend to be wet and humid, for instance, or dry; after all, there are regions that are very wet, others where it rains very little, and still others where it almost never rains, such as Egypt, according to Herodotus. It is best to sow in dry weather, before the rains have come. Here in Spain, when water does not run in the arroyos (ditches) until January, it is considered a good year for sowing. Virgil says, "*Humida folstitia hyemes optate ferenas agricolae*," meaning that climate conditions are ideal if the winter is moderately humid and dry and the summer is humid and wet. Virgil indicates that almond trees being heavy with nuts is another sign of a good year. With regard to planting in regions that are humid and wet, higher elevations and hillsides should be sown; and in those that are dry, plains and valleys should be planted. If a region tends to be cloudy, higher elevations should be planted. Of course, to enhance the probability of a good crop, one can sow everywhere, including remote locations.

Advice on sowing from other writers likewise pertains less to celestial signs and more to climate and soil conditions. Crecentino states it is good to plant when the soil is moist and the temperature fairly warm, because heat and water sprout seeds. Palladius confirms this assessment, and experience has proven its validity. If possible, one should avoid planting when it is very cold or freezing, or when there is a bitterly cold northerly wind that has originated in intemperate regions. Pliny asserts that the cold naturally closes the pores of seeds and soil, hardening them against the penetration of warmth and thus thwarting growth. Often when seeds are in the ground with moisture but little warmth, they will rot, or exhibit uneven and stunted development. Columella too advises that one should neither plow nor plant fifteen days before or after winter ends, due to the damaging effects of extremely cold weather. In warmer areas, sowing is advantageous after it has rained and the ground is soft, for weeds usually do not grow at that time.

Farmers must be conscious of the fact that where lands are cold and humid with thin soil, sowing should be done earlier, even in autumn, before the rains and extreme cold set in, allowing the seeds to sprout and grow, if only minimally. In regions that are much drier and with warmer richer soil, sowing should be done in mid-winter since these lands can better tolerate cold temperatures and heavier rains. Crecentino twice mentions that in very wet, cold, and humid areas, sowing should be done earlier, and in warm and dry places later. This point is also raised in the writings of Columella, Pliny, and Marcus Cato. If the soil is dark and fertile, sowing can be delayed somewhat, since darker soils, according to Theophrastus, are typically more capable of tolerating heat and retaining moisture.

In any case, sowing should never be postponed until December, the month agriculturalists designate as "*abruma*" ("time of frost"). It is said that seeds sown prior to this month take no more than seven days to sprout, while those planted during or after this month take about forty days. Pliny and Marcus Varro confirm this, and cite signs in the heavens that indicate when sowing should begin. Virgil recommends that wheat, barley, and other similar seeds be sown after the setting of the Pleiades (called *las cabrillas* in Spanish)—which would be immediately after mid-October, according to Columella, quoting from Virgil. This sign, however, is very difficult for farmers to understand and recognize. Moreover, I do not believe this rule is accurate for all regions, since it is determined by the movement of the heavens and the position of the earth, with constellations rising and setting in some regions before others.

Perhaps another rule of Pliny's would be more valuable to all farmers and applicable to any type of soil or region: instead of being guided by the heavens, sowing should commence when trees begin to lose their leaves. This rule applies to each individual region, is more predictable, and reflects the fact that the soil and trees prepare themselves for cold. Pliny further says that while gauging time for early sowing may at times be deceiving, planting late is always unreliable. Columella agrees, stating that sowing late is bad and should only be done as a necessity. Early sowing almost invariably results in good productivity, although this cannot always be correctly assessed when a crop turns out to be deficient for other reasons and a farmer attempts to remedy the situation by sowing later the following year. He asserts that one should sow after it rains heavily in the early fall. If the rains come late, one should sow even if it is dry, because seeds planted under dry conditions are as secure in the ground as in a granary; then, when the rains arrive, they will sprout and grow as well as if they had been recently planted. For this reason, it is advantageous if the soil is well worked, very loose and spongy, and maintained as such, to preserve the moisture it receives. Theophrastus confirms that to conserve moisture it receives, the soil should be well plowed numerous times. Also, soil that is well worked remains very easy to plow, responding almost as if it had rained.

Now that we have discussed the time to sow, we will focus on sowing techniques. Pliny recommends dispersing the seed evenly throughout the area to be cultivated. The furrow should be deep and the seed well covered, since seeds left uncovered will be wasted. Theophrastus suggests that it is better to plant less seed and cover it well than to use more and not cover it sufficiently. In addition, the deeper the furrow, the deeper the roots, and thus the more moisture they can absorb. Those who sow hillsides should plow as deeply as possible to ensure the seed is well covered so it won't be carried off by water, like the topsoil often is in such places.

Further, all agricultural scholars agree that it is essential to know the quantity of seed best suited to the soil. Not even Columella, however, thinks a specific measurement for seed can be firmly established. According to Crecentino, the quantity of seed is determined by the quality of the soil. In thin soil, seeds should be sown more sparsely than in soil of greater substance. In soils rich with substance, more seed should be sown, but not so many that the plants choke one another as they grow. It is important to maintain a balance so that no soil is oversown with more seeds than it can reasonably sustain. While overly used soils require more seeds, particularly if sown late, sparsely sown lands have a greater area for roots to grow larger and sprout more shoots. As for the different varieties of wheat, white wheat and *deraspado* wheat should not be sown as densely as brown wheat, because they tend to produce more shoots.

In general, Pliny suggests that when planting early, seeds should be sown more densely, and when planting late they should be spread more sparsely. Columella and Crecentino disagree, stating that when planting early seeds should be sown sparsely, and more densely when planting late. None of these authors, however, provides sufficient reasons to support his contentions, although Theophrastus gives a sound reason to support the latter view: when planting early, he asserts, seeds should be sown more sparsely, allowing them to root heavily and produce many shoots, and when planting late, seeds should be sown more profusely since many of them will be lost due to heavy rains and frost.

THE MOST SUITABLE SEEDS FOR PARTICULAR SOILS

For productive cultivation, farmers must be able to match seeds to the soils best suited for them. For example, the different varieties of WHEAT do not all grow well in the same type of soil, and consequently types of wheat found in some regions are unknown in others. Brown wheat (*trechel*), or golden wheat (*rubión*), which is the heaviest and most substantial of all types, grows better in the warm, rich soils of the plains than in humid or shaded lands, since the grain itself is naturally somewhat moist. If this type of wheat is planted in soils that are too cold and thin, the first time it grows it produces inferior grain; the second time, rye. Dark wheat (*arisprieto*), which is similar to ordinary brown wheat, is naturally adverse to humid and shaded areas, and requires rich soil and higher elevation. White wheat (*candeal*) tolerates thin, cold, shaded and humid lands better than brown wheat, and prefers moist, cold, and loose soil. Hence, this wheat is planted in some mountainous regions. I am not suggesting that it grows better there than in rich soils,

only that it generally does well in higher elevations. The *deraspado* variety of wheat (with thick bran) thrives best in higher elevations. Because this wheat has a thick husk with many layers, it is not easily damaged by frost and cold, unlike brown wheat, which has a very thin husk.

White and *deraspado* wheat have stiffer, tougher stalks and fewer leaves than brown wheat, whose heavy grains tend to fall and rot, making it better suited for humid and rainy lands. The same holds for spring wheat (*tremesino)*, which is of the same variety. Even so, according to Columella, spring wheat differs from the white and *deraspado* varieties because if sown in the fall, rather than at the beginning of summer, it grows better and more prolifically. He also indicates that spring, white, and *deraspado* wheats should be planted in the fall or at the beginning of winter. Moreover, the regions that are best suited for spring wheat are cold, foggy, and rainy, with heavy snows. After surviving the cold and humidity, it can better resist the heat of summer. This is sensible counsel, considering winter lasts longer in cold regions, and the heat and moisture of the brief summer aids the wheat's growth.

Regarding the best times to sow spring wheat, Columella recommends January in warm areas and February in colder regions. Palladius affirms that it is best to plant spring wheat in cold, humid, and snowy places, since it generally does not thrive in warm, dry lands. If this wheat is not sown in January, or February at the latest, which is undoubtedly the best time to sow, all these authors agree that the air becomes too warm, although it may be sufficiently humid. In tepid lands—which are neither hot nor cold, but humid and moderately warm—spring wheat should be planted in January and February, or at the beginning of March in colder regions, but always before the equinox in mid-March. Columella affirms that spring wheat can be planted late if the lands are cold and humid. In addition, Pliny recommends that the soil for spring wheat be loose, thin, and spongy. Theophrastus concurs, stating that soils rich in the summer tend to dry out more easily with the sun than thin soils, particularly in times of insufficient rain. Rich soils require more water to ensure proper growth of plants. Thus, in dry years wheat invariably grows better in thin, loose soils than in hard soils because the former do not require as much moisture. Moreover, Theophrastus contends that since spring wheat has smaller, fewer roots, with very few shoots, it should be heavily planted. Pliny and Columella debate this point, Columella maintaining that spring wheat is not a separate variety, and Pliny asserting that it is, as verified by agriculturalists, and one most suitable for cold, snowy lands.

Most agriculturalists agree that BARLEY thrives in loose, airy soils, although Crecentino differs, stating that barley should be planted in rich, albeit loose, and dry soils, because it can be destroyed by excessive moisture. This seed produces a hollow, soft, and thin stalk with many leaves, and if it falls due to excess foliage, it is unable to spring back up like wheat and could rot. Thus, it should not be sown in humid lands, where there is much snow, dew, and cloud cover. Such regions are more unsuitable for barley than for any type of wheat because its grain, according to Theophrastus, is almost completely uncovered. For this reason, it is best to sow barley in higher elevations and windy places where there is less humidity and dew than on plains and in valleys, where the wind carries off any moisture on its stem. If either the sun or warm wind that blows intermittently heats the dew on the stem, it will burn and dry, resulting in a condition called *rubigo* in Latin, and in Spanish, *nublo* (plant fungus). Crecentino states that it is best to plant barley in valleys. Theophrastus, however, recommends sowing it in higher elevations, because the damage caused by humidity is greater in the valleys and plains. I firmly believe that barley sown for grain can be planted in valleys as long as they are windy and cloudy. Barley should not be sown in mire but rather in dry soil, because it will perish in mud—unlike wheat, which can thrive in muddy soil and in fact retain greater heat, enabling it to withstand the cold and become more resilient for growth.

There are at least five types of barley, each composed of several varieties. First, there is common barley, called *ladilla,* with a spike similar to that of wheat. It grows in warm lands and its grain is much fuller and heavier than the type that Columella calls *"diftica,"* or *"galaica,"* which thrives in cold and humid regions. In addition, there is a type that thrives in warmer climates, which Columella calls *"cantierina."* Crecentino affirms barley is not a cold grain and that all types of barley sprout multiple shoots, with some varieties growing better than others. According to Theophrastus, barley has an extensive root system and thus, if sown for grain, should not be planted densely as it tends to grow fast, particularly when sown early. When sown later, additional seed is required, as

some may be lost. If planted for forage, on the other hand, it should be sown much more densely. Barley should always be sown in soft, well-plowed soil, allowing the root system to grow deep in the land.

If the soils to be sown are varied, with some being compact and other loose, and if the land consists of valleys, plains, and mountains, Theophrastus suggests planting wheat in the valleys because it requires soil of greater substance. Barley should be sown in moderate-grade soil in accordance with the aforementioned recommendations, ensuring that the land is not shaded, humid, or dewy. Pliny explains that the best type of this grain is white barley, since its color indicates it is not wet. Further, the white barley seed should have a pleasant odor, be heavy and full-grained, and neither wrinkled nor moist. Barley that is moist and releasing an unpleasant odor is generally sterile and will seldom sprout.

A fifth type of barley, called *ordiate,* grows without a hull, much like wheat. It should be prepared for sowing as follows. Soak well-selected barley for one or two hours, and sieve it through a sack or rough fabric until the hull is removed and the grain is clean. Then if sown in good soil or along the edge of a garden or orchard so that it can be easily irrigated, it will grow there perpetually. It is excellent for medicinal remedies when boiled in water.

RYE should be planted in temperate climates in warm rather than cold regions, as it tends to spike early and easily freeze. In fact, the best rye fields are found in hot or temperate regions. Theophrastus recommends sowing rye in light soil because of this grain's hollow, thin stem. Moreover, it can be sown in high elevations not generally suitable for wheat. Rye sprouts many shoots; therefore, it is best to sow it lightly, unless it is being grown for fodder. Also, because its seed is small and incapable of resisting cold, rye must be sown early. If sown late, it can choke and rot due to cold, rains, snows, and frost.

In addition to seeds of wheat, barley, and rye, a myriad of others proliferate. These varieties of grain correspond not only with specific geographic and climate conditions but also with unique nutritional needs. It is through the divine providence of God that there are appropriate seeds for each type of land, thus providing for human and animal sustenance.

THE BEST TIMES AND METHODS FOR WEEDING CULTIVATED AREAS

To weed *(sallar)* entails clearing both damaging and worthless weeds from the land where plants have been sown. As our Lord says, If the bad exceeds the good, it will not allow the good to thrive. Just as well-governed communities banish the bad and corrupt so they will not harm the good and virtuous, the same should be envisioned with plants. Dead branches must be removed from trees, sickly livestock eliminated from herds, and weeds eradicated from fields of grain and other crops. In fact, weeding is so essential in some regions that if it is neglected, particularly during rainy years, crops will be choked.

Even though some ancient agriculturalists maintain that weeding is unnecessary, if not outright damaging because it uncovers root systems, I disagree. There are some places that can go without weeding and still produce good grain, but when weeded, the soil becomes porous, soft, and spongy, and the root systems produce more shoots, better spikes, and bigger, heavier, and more nutritious grain. Weeding is also important because it produces straw that is softer and tastier for animals. In rich, fertile lands, weeding is especially important, since more weeds tend to grow on these lands, especially during rainy years. Regardless of time or region, weeding is invariably beneficial provided it is done at appropriate times and using methods that will not damage crops.

Weeds can be pulled out, but in many instances to minimize damage to roots of other plants, weeds are removed with a tool agriculturalists call an "*almocafe*," a Moorish word I will refer to as a hoe. Agriculturalists agree that wheat should not be weeded or hoed until it has produced four leaves, since only then is the root system sufficiently developed to avert endangerment through weeding or hoeing, and the sparse leaves do not impede the process. Barley should not be weeded or hoed until it has five leaves. Both wheat and barley should be hoed after they have rooted and before they sprout shoots, because after sprouting they can be more easily stepped on and broken, and thus die. Also, weeding should be done on calm days, when there is no possibility of frost, because if the wind or cold gets to uncovered roots, they will dry out or be damaged. Columella adds that in warm regions weeding should be done before the bitter cold sets in, around December or shortly before, while in cold places it should be done in the spring, around February or before shoots begin to sprout. After hoeing, always ensure that the roots are well covered.

If the farmer does not have time to remove all weeds, he should at least eliminate those with extensive root systems that take up considerable space, such as thistles, hollyhock, darnel grass, downy chamomile, and chicory. Virgil says, "*El amaris intima fibris.*" ("Remove all weeds that have deep

roots and produce branches.") Above all, poppies should be eliminated because they take up too much space, absorb all the moisture, burn other plants, and usurp the substance of the soil. As Virgil states, "*Urunt lethaeo perfusa papavera fomno.*"

According to Columella, barley should be weeded when the land is dry, in order to avoid mistakenly cutting entire root systems, which would cause plants to become mangy. Pliny confirms this fact and adds that if wheat or barley has too much growth, ensure that the roots are well-covered and the plant is not, for then it would rot; a small plant with only five or six leaves, however, will not be damaged, even if it is covered completely.

There is yet another method for killing weeds, which Pliny calls "*lirar*" and which is used in some parts of Castile (where it is referred to as *arejacar*) for wheat and barley, but it is implemented only when plants have developed a good root system, in the following manner. If there are small weeds remaining after the bigger ones have been pulled manually, the soil can be plowed in the opposite direction from how it was sown. While it might seem that the grain would be ruined by uncovering some of its roots, in fact this procedure moves the soil in such a way that root systems grow more extensively, resulting in more shoots and bigger and more numerous spikes. Thus, if one plant is lost, more and better plants will grow. This method of weed eradication is not appropriate for all lands. It is not recommended for sandy soils because the roots are too readily uncovered, and is best used on good, rich lands, even if the soil is loose and sticky. In addition, it should be undertaken when the weather is cloudy, humid, and rain is expected—not when it is dry, sunny, or windy, particularly during dry winds, or when it is cold and frosty. A similar procedure can be used when it has rained so heavily that plants recently sown cannot grow due to the hardness of the soil's surface. To save the crop, the land should be plowed again, in the opposite direction from the previous furrows. The soil plowed from underneath covers the hard topsoil, softening it in the process. This should be done when the weather is humid, the plants small, and the sun not shining intensely.

I know of no agriculturalist who suggests weeding rye. For one thing, rye matures very quickly, outgrowing any weed and choking it. For another, rye has an extensive root system that produces many shoots, thereby preventing weeds from growing. If for some reason rye requires weeding, the same indications for weeding other grains should be followed.

In some instances, such as when plants grow quickly and frilled and are too moist and ripen too rapidly, causing them to fall and rot, grazing the land can be beneficial in controlling plant growth. Some suggest reaping it instead, because livestock trample the land excessively, tearing out roots, and also tend to not graze evenly, so it may have to be reaped anyway to make it all even. Generally, however, most agriculturalists agree that permitting livestock to graze on land is preferable to reaping it. Pliny, for one, states that if livestock graze, the spike will not breed worms or other diseases.

Even so, no author whose works I have read addresses methods for grazing livestock on the field, when it should be done, or which animals to use. Based on common sense, it seems it would be best to graze large livestock like cattle, horses, and donkeys, because they tend to eat leaves, not pull out roots, and cover considerable ground. Shackled animals should not be grazed, because when they move they tend to jump and dig deeply into the sown soil and since it is difficult for them to move, they will eat everything in a certain area, including the roots, before moving on, leaving the land devastated. The worst livestock to graze are oxen, because they uproot too many plants. It is good to graze hogs on sown fields, particularly if they are hungry. They should be moved rather quickly, however, because if left in the field after they are sated, they begin digging with their snouts, causing much damage. If sheep are allowed to graze, they must be moved almost continually so that they eat only the tops of the plants. It is better if they are driven through the field two or three times instead of just once. Small livestock, particularly sheep, get bloated quickly and can perish from overeating. In general, animals should be grazed on sown fields before the plants become straw or develop shoots. Because shoots often grow fuller than the plant itself, animals tend to graze on them first, while the stem proceeds to rot. It is important to graze animals on barley first, given that it is very tender and rots quite easily, unlike wheat, which has a strong stalk.

On the other hand, crops must be protected from some birds and animals. Cranes and domesticated geese should be driven away because their sharp beaks cause damage to vegetation and because they eat entire plants and uproot small plants completely. Further, the excrement of all waterfowl, including geese, is very damaging to grain and vegetable fields, desiccating and burning everything it touches, as if hot embers or boiling water had been poured on the plants. It is even more imperative to drive away goats, which run in large herds and move quickly, while domesticated geese seldom leave their villages. Finally, if the land is irrigated, it should be safeguarded from horses so that they do not trample it and thus severely damage the plants.

Another means of protecting crops is the use of fences. Ideally, fences for grain fields should have high poles with a deep, wide exterior trench and not be made of brambles so they become a natural home for rabbits that cause considerable damage to plants, especially young and tender ones. In Portugal and many parts of Castile, if the sown fields are near a road, a wide section of lupine is planted between the two, dissuading animals from entering the field to graze. Lupine is beneficial for oxen, however, and for purposes beyond framing an enclosure. Buckthorn, too, can be utilitarian, since rabbits rarely make their homes there, but unfortunately it is often inhabited by sparrows, birds that cause the most damage to spikes, particularly those of barley. Since grains of barley are almost uncovered, when sparrows land on the spikes, they eat a grain or two and throw the rest to the ground with their wings, leaving only straw at harvest time. Upon noticing the impact of sparrows on his barley, one farmer reaped the straw and plowed the field full of grain, thus growing good barley the following year. Essentially, however, it is best to sow barley far from villages, in areas free of trees or plants where these birds can hide, as they tend not to stray far from villages for fear of hawks.

Having used all these methods for protecting crops and ensuring their proper growth. nothing else needs to be considered until harvest time. Meanwhile, if the farmer is interested and wants to remain busy and

content, he can learn about other agricultural endeavors, such as vineyards and orchards. Moreover, he should always pray to God for good weather and health.

Effective Reaping, Threshing, and Storing

The reaping, threshing, and storing of grains must be done quickly and properly if the entire year's endeavor is to be preserved. With that in mind, certain tasks must be completed in advance of the harvest. Then the harvest, the threshing of the grain, must be approached with care. Appropriate storage follows so the grain can be well preserved till the next planting season and beyond.

BEFORE THE HARVEST

It is essential to level and prepare the threshing ground, the place where the grain will be separated from the chaff, before reaping so that later the harvested grain can be brought directly there rather than becoming lost in the stubble. If threshing grounds are near each other, people can assist one another as needed, which is particularly helpful in the event of strong winds. Unthreshed piles should not be placed close together, however, in case one catches fire.

Laboring in a cold, dry place with naturally occurring winds helps separate the chaff from the grain. But Crecentino advises that threshing grounds should be protected from southwesterly winds, particularly if the region is humid, because such winds are moist and quickly spoil grains, causing them to grow grubs. Also, threshing grounds should be located far from vineyards and gardens, because the chaff, although a good fertilizer when applied to roots, causes damage when settling on plant leaves, breaking them and increasing the likelihood that the fruit, especially grapes, will become infested with worms. To better preserve the wheat, threshing grounds should likewise be located away from places with bad odors.

The area surrounding the threshing ground should be clean so the straw or chaff that falls on it will be easily gathered and safe to eat. Regarding construction of the threshing ground itself, Varro says it should be round, elevated in the center, and with sloping sides so the water runs off in the event of rain. Rock lining will further facilitate the threshing due both to its hardness and its ability to prevent moles, rats, and ants from digging holes and damaging the area, thus ensuring the grain will come out cleaner, free of pebbles, dirt, and other impurities. Columella states that the cleaner the wheat, the longer it will be preserved and the healthier it will be for consumers. If the threshing ground is to remain dirt, however, the soil should be well sieved; drenched with regular water mixed with olive water; firmly padded or pounded down by people, livestock, or with a hammer; then turned and soaked with olive water, which hardens the ground, prevents weeds from growing, and as Varro notes, poisons ants, moles, and similar rodents.

If possible, the threshing ground should be located in a place with heavy, claylike soil so that when it is soaked with the olive water it obtains an almost impermeable surface. By contrast, if the area is sandy, the grain will become mixed with sand, which is obviously detrimental.

Another aspect of preparing the threshing ground is providing for possibly harmful weather conditions. To protect the grain from unexpected heavy rains that would cause great damage, it should be piled in a temporary structure or covered by tarps. Moreover, unthreshed wheat should be piled either in round or long stacks so it can be quickly covered. If stacked on dirt and exposed to the elements, unthreshed wheat should be gathered tightly into sheaves and tied with vine shoots, keeping the spikes concealed as much as possible within the sheaves to protect the grain from both rain and livestock; then the sheaves can be covered with an impermeable material for further protection, and stored this way for a period of time.

Before beginning the harvest, the farmer should also prepare ties. While many individuals use spartium for these, others use willow twigs, and still others plant rye within the wheat field because its stalks are long and strong enough to be effectively used as ties. Although the latter method can be useful if the rye is old, in my opinion it is not appropriate to sow other types of seeds with wheat or barley. It is better to sow the rye separately so it can be easily reaped for use as ties and its grain shaken off but not mixed with the wheat or barley, and thus not diluting the taste of these grains or reducing their nutritional value. In fact, the rye should be kept separate even from straw, which also can be harmed during threshing. The best ties are those of *valago* (bass weed) in places where it is plentiful. To improve their endurance, they should be well moistened before use.

WHEN TO HARVEST

The dangers of harvesting are similar to those of sowing, where heavy rains and other harmful occurrences can damage the crop if preventive procedures are not undertaken quickly and properly. If the grain becomes wet, it will rot and acquire a stench, resulting in unhealthy and unpleasant-tasting food that cannot be preserved for any length of time. Even straw acquires a stench, and livestock that do not refuse to eat it may contract glanders or other diseases. Therefore, while harvesting, it is best to complete the work quickly, without leaving tasks for another day.

While harvesting wheat, one should not wait for it to become exceedingly dry, since the stalk is thick, the grain well covered, and according to Columella, as the summer gets increasingly warmer, unharvested wheat falls and decays. At this time of year, hail and heavy rainstorms are frequent occurrences as well, potentially endangering the grain. Excessive sun saps its substance, a development referred to by Latins as *siderare*, meaning to burn with the sun's heat during the sign of the Dog Star. Pliny says that the more rapidly wheat

is harvested—beginning, ideally, as soon as it is dry—the fuller and more improved the grain, and the better its chances of being preserved. Generally, agriculturalists concur that harvesting should be done earlier rather than later to make wheat less susceptible to damage or disease. If good weather is a certainty, harvesting should be undertaken when the moon is waning rather than waxing and preferably at the end of the day of a waning moon, as it is ideal for preserving the grain and makes it less susceptible to moisture-based infestations such as grubs.

Barley must be harvested even more quickly, since it ripens earlier and the grain is naturally less covered, causing it to dry quickly and fall off its spike. Hence, if it is sunny, barley is best harvested in the morning. During a full moon, it can be harvested at night, as is done in extremely hot parts of Andalucía and Africa. Some agriculturalists recommend that the bundles remain spread out on the ground for three or four days after harvesting, claiming that this fattens the grain; but since the barley grain is dry, upon absorbing moisture from both the soil and the night air, it swells and becomes soft. This will not damage the grain for immediate use but can be harmful if it is stored for later use or sowing.

THRESHING INSTRUCTIONS

It is best to thresh on sunny days, because heat and dryness facilitate separation of the grain from the chaff. If threshing must be completed rapidly, the separated grain should be covered, while the remainder is being threshed. The reason for haste is that if the unthreshed wheat becomes wet, it will be more difficult to separate the grain from both the spike and the chaff.

There are two main methods for threshing. The first is to use spiked implements known as harrows to pulverize the grain. A much better method is to use livestock, particularly horses, because they quickly and efficiently crumble the chaff and break up the spikes. To prevent them from choking, tie their headstalls together rather than their necks. The best horse should be the leader, with the others following. An area in the center of the grain pile should permit the person guiding the animals to be elevated. Alternatively, some farmers have a person on horseback guide the animals, while others place a firm, round, heavy beam in the center of the grain pile with a large ring around it that can turn easily, then tie a rope from the ring to the lead horse. In both instances, a winnowing fork is used to stir the heap of unthreshed grain, dispersing it under the animals' hooves.

Whether using harrows, horses, or another method for threshing, the grain should be dry, facilitating its separation from the spikes. Pliny affirms that animals, especially oxen, prefer more finely threshed straw. All authors agree that after the grain has been threshed and cleaned, it should be left on the threshing ground for a day, allowing it to cool off, which according to Pliny, helps to preserve it. To accelerate the cooling, the grain should be divided into two or three piles and stirred with a shovel at night or whenever air temperatures have dropped.

GRANARIES, SILOS, AND OTHER STORAGE FACILITIES

Granaries and silos are both suitable for storing wheat. Barley and rye, however, should not be stored in silos, because these grains are very dry and will spoil and acquire a stench in storage if there is any moisture. These grains, and wheat as well, are best stored in granaries that are high, cool, and well ventilated, far from moisture, stables, and other places with bad odors, and should have small windows facing cold northerly winds. According to Columella, and as experience has demonstrated, such winds originate in a part of the heavens that is very cold and rather dry, a combination of conditions that preserves grain in a granary for a long time. Granaries should not have southwesterly ventilation, however, because this air is warm and humid, which quickly spoils the grain. Pliny and Theophrastus both recommend that granaries not be whitewashed because lime is hot and could spoil the wheat.

Granaries should have very sound floors and walls and the roofs should be free of leaks, to prevent mice or grubs from entering. Agriculturalists agree, that to safeguard stored crops against worms, grubs, and other vermin, the walls and floors of granaries should be covered with mud mixed with unsalted olive water and green leaves from wild olive trees, or if no wild olive trees are available, leaves from ordinary olive trees. This process should be repeated, as necessary, to keep the walls and floor solid. Pliny affirms that where there is no air, there are no grubs, and I concur. To prevent air from penetrating, a granary should be constructed like a vault, which also affords greater protection against fire. In fact, some individuals build granaries so tightly that they must be filled through a hole on top and the grain removed through a hole at the bottom. To ensure that all the wheat can be removed, the floor is slanted toward the hole.

Palladius adds that neither mice nor ants will penetrate granaries plastered with mud mixed with fresh cow dung. Varro suggests that wheat in the granary be sprayed with olive water, but I am reluctant to endorse this method since it could give the wheat a bad taste and adversely affect its use as a seed. Columella recommends that if wheat begins to grow grubs it should not be shoveled, as others suggest, because grubs grow only on the surface of piles and will not penetrate deeper than the width of the palm of a hand. If the wheat is

shoveled, it will mix the bad with the good and ruin it all, leaving only the possibility of selling it or turning it into flour. If there are grubs and weevils, turpentine tree leaves or juniper branches should be placed among the wheat, causing the vermin to flee or die. Pliny claims that barley does not breed grubs because it has a thinner covering than wheat. In addition, barley is cold, as are rye and oats, which also do not produce grubs. Wheat, on the other hand, is hot, and thus more susceptible to spoilage. Finally, granaries should have numerous compartments where different varieties of seeds can be stored and, if possible, separate compartments for new and old seeds.

Silos in very dry places are beneficial for storing grains other than barley and rye. Varro instructs that silos be located in extremely dry places with very dry, hard, claylike soil, in higher elevations free of humidity and moisture. Straw should be scattered throughout to prevent the grain from absorbing the smell of the soil. Rye is preferable for this purpose because it is cold. Varro agrees, suggesting that wheat stored in this manner will be well preserved for fifty years, and millet for over a hundred.

Further, some varieties of wheat can be stored more successfully than others, and also certain growing conditions contribute to successful storage. Brown wheat is better for storing than white wheat, due to the robustness of its grain. Wheat grown at higher elevations and without fertilizer is better for storage than fertilized wheat grown in lowlands. Wheat grown in dry soil is better than grain that was irrigated. Also, wheat harvested during a waning moon is better than wheat harvested during a waxing moon. All these generalizations pertain to the storage of large quantities of grain. When there is only a small amount of grain, it can be stored in a large earthen jar, which will preserve it as well as any granary in the world. Palladius claims that wheat can be stored for a longer time if an herb called *coniza* (great flea-bane) is added generously to the grain, mixing it throughout. In Spanish, this herb is called *ojo de buey* (ox-eye), although I am not familiar with it. In my opinion, the safest method for wheat storage is to grind it into flour when the weather is cold and store it in large earthen jars mixed with unground salt, which can later be removed by sieving the flour. While this method is effective for storing flour, the most delicious bread is of course made from fresh flour.

About Grains

WHEAT

By nature, wheat is moist and warm, and if eaten raw, it will produce thick phlegm and be detrimental to digestion. It causes inflammation, gastrointestinal pain, gas, diseases, and it lacks nutritional value. Toasted wheat is better nourishment, particularly for those who retain water or are phlegmatic, since it tends to dry and does not cause gas but rather settles the stomach. White bread made from flour is very healthy and tasty, particularly when it is well kneaded and well cooked. For people who are depressed, however, it is harmful. Wheat bread is better than any other type of bread, as it most closely approximates the human constitution. Because wheat produces viscous bodily fluids, salt is usually added to remove superfluous fluids. For individuals engaged in hard labor, bread without salt is better because their bodies are always warm enough to digest it without generating excessive fluids.

Wheat, prepared in various ways, is not only nutritious but also curative. According to Crecentino, wheat that has been washed well with hot water, had the husks removed, then been cooked with almond milk and honey or sugar helps clear the lungs of viscous fluids, produces healthy blood, and is excellent nourishment. Wheat cooked in wine and water and placed on a woman's breasts will alleviate the hardening caused by her curdling milk.

Saint Isidore explains that wheat flour mixed with honey is good for scabs and facial impetigo. If mixed with honey and hog lard and placed on women's breasts, it reduces swelling or alleviates tense and swollen nerves and muscles. But if eaten regularly, he cautions, wheat flour can produce bladder and kidney stones, due to its viscosity. Nonetheless, the more sugar added to cooked wheat, the better, making it less damaging and preventing it from obstructing the liver and hardening the spleen.

In addition, Pliny claims that those suffering from severe gout (*gota*)—not to be confused with epilepsy (*gota corral*) but rather the type afflicting the feet and hands (also called *podraga* or *chiragra*)—should place the affected area in a pile of wheat. He also suggests that if someone has frostbite and is numb, wheat oil taken with a hot poker should be rubbed on the affected area. This method is also beneficial for impetigo and fetter.

Specifically, white wheat is helpful for people who retain fluids and suffer from bloating, since it is a phlegmatic because it is dry and absorbs excess fluids. It is also a good cure for head colds and runny noses, or flu. A remedy commonly known as *almidón* (starch), made from white wheat, is beneficial for those who suffer from consumption, because it naturally heals lesions in the chest. To make this remedy, prepare the wheat as follows. Soak clean wheat for eight days in water, changing the water four or five times a day. After the eighth day, remove the wheat from the water, place it in a strong cloth sack, put the filled sack into a press, squeezing the milk into a clean earthen jar, then spread it out on a kneading trough to dry. Lightly covering the kneading trough with a little yeast in advance will make the remedy whiter, harder, and less likely to sour. Stewed with almond milk and sugar, it is very similar to a dish made from milk, sugar, and rice flour (*manjar blanco*). Over all, *almidón* is excellent for curing illnesses, highly nourishing, and easily digested.

Brown wheat, which is moister and colder than white wheat, lowers the fever caused by cholera. Brown wheat is also better suited to the summer rather than the winter, and to humid climates rather than dry, hot weather. Further, it is said that if wheat grain is roasted and ground, and spread on the bite of a rabid dog, it will help considerably. The same is true if it is applied to a variety of inflammations and boils. If wheat is simply boiled in water and eaten, as is widely done, it is harmful to digestion because it causes intestinal gas and swelling, producing thick bodily fluids, opilations, worms, intestinal rawness, and colic. When boiled and stewed, with salt and honey or sugar, it is very good for diarrhea. A cataplasm made from brown wheat and rose oil or salve softens abscesses, relieves the pain resulting from them, and cures them more rapidly.

Finally, wheat bran boiled in strong vinegar and spread hot over leprous sores will dry them. Wheat bran boiled in rue juice and placed on a woman's breasts after a pregnancy softens them and reduces the

swelling. The same paste also alleviates snakebites and poisonous insect bites and reduces diarrhea.

BARLEY

Barley can be used in many ways to treat a variety of ailments. Extremely cold by nature, it is given to people who suffer from fevers, including fevers caused by blood diseases. It is also good for cleansing the body internally. Barley water purifies and refreshes, effectively treating illnesses caused by heat and by agitation of the blood.

Barley's external applications are wide-ranging. Individuals afflicted with ringworm should boil barley in water and spread it on the affected area while hot. Those who suffer from lumps, swelling, or abscesses, if not hot, should boil barley and bran in water until it attains a paste-like consistency and apply it to the affected area. To alleviate the effects of gout on the limbs, cook the barley in vinegar, add some quince, and apply it to the affected limbs.

Taken internally, barley water is an excellent pectoral and thus a good treatment for pulmonary illnesses and excellent for consumption. Drinking fennel boiled in barley water increases the production of milk in nursing mothers. Barley boiled with *coronilla del Rey* (a type of sweet clover) and garden poppies produces an effective treatment for backaches. Beverages made from barley grain or barley flour should be brewed as follows. Clean the barley, shake it to remove its husk, and wash it repeatedly; then boil and squeeze it to extract the liquid, and mix it with sugar. Barley water should be taken sparingly, because in increased amounts it adversely affects the stomach, increasing water retention. Finally, barley bread is highly nutritious and particularly beneficial for cholerics, as it is very easily digested and produces healthy blood. Crecentino notes all these benefits, and Pliny adds that people who eat barley bread will not suffer from foot problems, since it is a softening agent that refreshes this area.

RYE

Rye produces bad bread that is heavy, moist, and viscous. While unhealthy for those whose stomachs are unaccustomed to it, rye is less harmful for those who exercise frequently. It is also less harmful, and more flavorful, if a lot of salt is added to it. The negative aspects of rye can be countered as well by mixing a one-third rye flour with two-thirds white wheat flour or *deraspado* wheat flour to make bread.

Nevertheless, rye flour is excellent for *empanadas* (filled turnovers), since it is cold and protects the filling from spoiling. Before grinding, it should be placed in the hot sun or in a hot oven to eliminate some of the excess moisture. Rye flour is also good for fattening oxen and hogs and, when cooked, for fattening horses and mules, although it should not be given to these animals in an excessive quantity.

Regarding its medicinal value, Pliny states that rye flour boiled to an almost pastelike consistency and eaten hot is an excellent remedy for those with diseases that cause them to spit up blood, since it is a coagulant that heals bleeding. I believe that if it heals externally, it must also heal internally.

STRAW

Straw is a continuous source of sustenance for the animals we value. Therefore, we should ensure that it is properly preserved and stored so they are fed healthy food.

Straw should be stored in haylofts and protected from dust and moisture. Also, to maintain freshness it is best to empty all the old straw before filling the hayloft with the fresh straw. If straw becomes wet, rinse it in the threshing area and stir it with a shovel so the sun and air will penetrate it. Wet straw is not liked by oxen and other livestock, and causes diseases. It is wise to conserve more than enough straw to feed livestock during what could be a long and bitter winter, since it is preferable to have too much rather than not enough. Storing sufficient straw saves one the trouble of finding or buying straw at a time when it may be difficult to access or expensive.

According to Pliny, barley straw is tender, and thus preferred by livestock and oxen over wheat straw, which is dry and hot and thus better during winter. Most animals do not eat wheat straw as well as barley straw because the former is rougher, although it has more substance. Oxen, however, will eat wheat straw, especially if it is well threshed. Further, brown wheat straw is more tender than white wheat straw, or straw of the *deraspado* or spring wheat varieties. Rye straw is cold and causes painful stomachaches, so if it is given as feed to livestock they should not drink for a while after eating but instead drink before they eat.

To give the straw flavor and make it more appealing to livestock that do not readily eat straw, Pliny recommends spraying it with brine and mixing it while it is still in the threshing area (after the wheat has been removed), then drying it before storing it in the hayloft. Working animals should be fed less than animals that do not work and should be fed only when they become quite hungry.

OATS

Vincencio marvels that very few reliable medical books mention oats despite their many medical benefits. In fact, among the ancient agriculturalists only Crecentino addresses when to plant oats and the appropriate soil, though he says nothing about how to sow the grain. He maintains that oats grow better in plains and valleys than in higher elevations, preferring cold rather than hot areas.

There are two types of oats. The first type, found in the mountains, is dark and downy with a thick, long stalk. The second type, which is planted like wheat and barley, has a white, smooth grain. According to Virgil, Pliny, and Theophrastus, because oats have a very deep root system that spreads widely, they impoverish the soil. Hence, soil in which oats grow should afterward lie fallow for a time. Since oats produce deep roots that spread widely, they should be planted very lightly in heavy soils that are loose and deeply plowed although they can also thrive in light soils. Oats should be sown like wheat, in October or November if the lands are dry and somewhat hot, but preferably in mid-February or early March, particularly if the lands are cold and humid. They should never be sown in mid-winter, however, because the extreme cold will choke them. Oats should be weeded like barley and do not require grazing, because they grow very quickly and there is no chance of their falling down and rotting. As a crop, oats should be protected from livestock because the grain is very tender and tasty, particularly the shoots, and having eaten it once the animals will crave it in the future. Moreover, oats are harvested at the same time as barley, and threshed and stored in the same manner.

Used as a remedy, oats are cold and dry and thus are given to people who have fevers, especially those with consumption. In addition, oat flour used as a medicinal paste softens hard lumps. Green oats generate healthy blood, and dry oats are good during the arid, hot weather of summer. Further, oats can be an excellent source of sustenance for livestock and are used throughout France and Germany to sustain horses and other animals. They are easily digested, and unlike with bran, the livestock neither become sated nor gain weight.

About Legumes

CHICKPEAS (GARBANZOS)

Chickpeas, although common, are detrimental to soil because they are salty, and salt sterilizes soil, and because when harvested they are uprooted, taking the best of the soil with them, according to Pliny and Crecentino. Everyone knows that salty lands are without redeeming value, even though their emaciated weeds are tasty and beneficial for livestock. In fact, there is a psalm that addresses this point, with the words "*Posuit terram eorum fructiferam in salsuginem.*" This is why when an enemy is defeated, or a village discovers a traitor, their lands are plowed and salt is scattered on them so that nothing will grow there, as punishment.

Although chickpeas, because of their salt content, may be more damaging to the land than any other legume, there are ways to minimize their saltiness. According to Palladius, soaking chickpeas in warm water that will not burn them a day before they are planted has two benefits: it dissipates their saltiness and it makes them sprout more quickly, grow larger, and be more tender. Pliny agrees with this recommendation, adding that chickpeas should never be planted without first soaking them. Virgil suggests soaking chickpeas, and in fact any legume, overnight in unsalted olive water mixed with saltpeter before planting them, maintaining that this makes them sprout better, enables them to grow more tender and tasty, and protects them from worms and other diseases. Theophrastus advises that if olive water is not available, saltpeter alone is sufficient. He also states that chickpeas and other legumes thus soaked should be planted in dry soils. I am inclined to think they would sprout more quickly, however, in soils that are fairly moist. If they are very hard, they should not be planted. If they are too tender, they lose the germ, and thus both the seed and the fruit. Fortunately this problem can be avoided by soaking them before planting, using one of the aforementioned methods to remove salt.

There are three varieties of chickpeas: white, golden, and dark. All varieties should be planted in the same manner. According to Crecentino, chickpeas thrive in temperate lands rather than in those that are either very hot or very cold—although chickpeas can thrive in most lands, as long as the soil is not exceedingly thin or depleted. If irrigating them, it is preferable to do so from below rather than from above, so the water does not touch the leaves. Irrigating from above carries salt from the leaves to the stalks and roots, thus killing the plant. Theophrastus adds that in many regions where chickpeas are grown, rain is very harmful to them because it washes salt from the leaves, which is then absorbed by the roots, causing the plant to turn yellow and dry. Theophrastus asserts that chickpeas will grow better, more tender, and tastier in thick, heavy, dry soils that are loose.

In these soils, chickpeas should be planted around the end of autumn. In moist soils, they should be sown after mid-February or in March, according to Palladius. If sown lightly, they will be able to extend their branches and grow better. Chickpeas should not be planted in heavily fertilized soils, unless the lands are not very moist or thin, and then only if the manure is quite old and decayed. The seed should be heavy and full, neither shriveled nor worm-ridden, and it should be dry but not old.

Columella states that these plants can be damaged if it rains while they are in bloom. To prevent such damage, plant them later, around March, April, or May, when showers will have already occurred and thus will not disturb the blooms. Theophrastus maintains that chickpeas require water only when sprouting. In addition, weeding is necessary only when the chickpeas are small, because as they grow they choke almost all the weeds. Pliny claims that the only weed to be cleared is *orabanche,* which I believe is *correhuela* (knotgrass) in Spanish—a weed that tightens around the plant, choking it. An additional way to prevent loss of chickpeas is to plant them away from roads and other well-traveled areas so hungry rabbits, jackrabbits, and other animals are less likely to be attracted to their salt content. In fact, when chickpeas are tender no one, not even a monk, can resist taking a handful as they pass. Shepherds and their cohorts are particularly unscrupulous in this respect, and if they are with women, no hail is more damaging. Therefore, it is best to sow chickpeas in well-enclosed or hidden areas so that no one knows the plants are there.

Regarding when to harvest chickpeas, Theophrastus notes that they ripen very quickly and should be harvested when they are ripe, dry, and, according to Crecentino, at the end of the moon's waning cycle. When transported to the threshing ground, they should be carefully wrapped in sheets, because otherwise they may slide off the vines. If the quantity is large, chickpeas should be threshed in the manner previously indicated with horses, because harrows are likely to break the grain. If the quantity is small, however, workers can thresh them with their feet or beat them off the vines with sticks. They should be cooled before storage, then placed in granaries to avoid spoiling in hot weather. Chickpeas for consumption are best stored in earthen jars that have previously stored oil or been moistened with olive water, while those for sowing should not touch oil. After the harvest, lands sown with chickpeas will be sapped of nutrients and should lie fallow for at least a year. A field that is to be replanted within the year should first be repeatedly plowed and fertilized with well-ripened manure.

Concerning the use of chickpeas as a remedy, one should know that they thicken blood and bile. Since these legumes are hot,

humid, and cause swelling, they are often fed to stallions that have to service many fillies. The white variety especially is effective for increasing milk production and reproductive potential, because they are short and moist. Chickpeas are also beneficial for the urinary tract and stimulate menstruation when it has been delayed. Further, they flush out kidney stones and break up bladder stones. According to Pliny and Crecentino, however, because of their salt content chickpeas are detrimental to open sores.

If one suffers from scabies, itchiness, or ringworm, warm chickpea broth applied to the affected areas cleanses them, particularly if made from the golden or dark types. Chickpea broth is better for treating medical problems than the grain itself because the broth contains all its best properties and medicinal values, as Hippocrates attests. To prepare the broth, first wash and soak the chickpeas, then boil them in the same water, since it contains most of their nutrients. This broth clears the voice, cleanses the lungs, and cures lumps and opilations of the spleen, liver, and gallbladder. According to Pliny, chickpea broth also cures epilepsy, a disorder caused by melancholy. Even though it increases heat, potentially causing putrefaction, Aristotle states that the broth reduces liver and spleen opilations produced by lack of heat. Further, Galen explains that chickpea broth stimulates urination and thus is beneficial for those with jaundice, purging the disease through the urine. It is also said that chickpea broth expels worms. Further, Dioscordis claims that cooking chickpeas with barley devoid of dust and chaff, kneading the mixture well, adding honey, and applying it as a topical paste cures head sores, cancers, and small wounds.

Finally, regarding uses for food, chickpeas are an excellent source of nourishment. Galen affirms that in many regions chickpea flour boiled in milk is good and full of nutrition. Green chickpeas are particularly good for cooking. And roasted chickpeas, unlike those prepared in other ways, do not cause swelling, but they are hard to digest.

BROAD BEANS (FABAS)

There are a great variety of broad beans (*fabas*). Some are large, others small, and all are either dark or white. Broad beans grow in all climates, hot and cold, but according to Theophrastus, those grown in cold regions can freeze and are not very tender when cooked after being dried. Consequently, the best broad beans are grown in hot or temperate lands. Broad beans prefer rich, sticky soils, in which they can grow very tender and large with thin skins; by contrast, when grown in thin, loose, sandy soils, they are small and hard. If rich, good soil is not available, well-ripened manure should be used as fertilizer. A negative aspect of this procedure, however, is that manure attracts lice and grubs that consume plants from the roots, especially broad bean and garlic plants.

Broad bean fields are best situated in breezy valleys, since such lands tend to be very wet and fertile. Plains or higher elevations are less suitable for broad beans, unless they are moist and sufficiently fertile. And while broad beans grow well in cloudy regions, they do not thrive in shaded areas. Nor will they do well if enclosed, for then they tend to be destroyed by lice.

There are two appropriate times to sow broad beans: either around October or November, or from mid-January through the end of February. It is generally best to

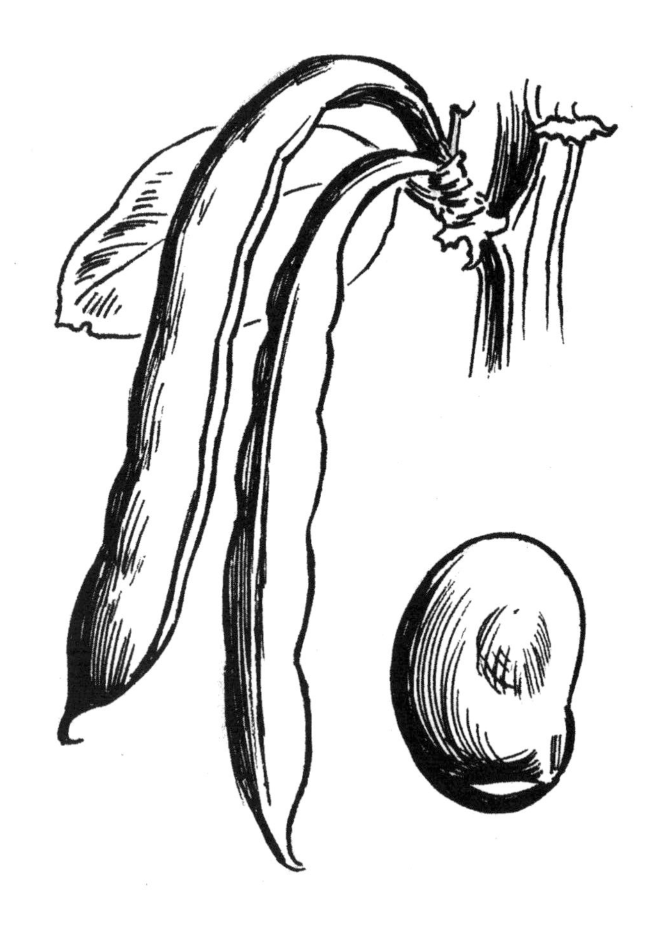

sow them early, because they take a long time to sprout and can be severely damaged if exposed to freezing temperatures before sprouting. Theophrastus maintains that good broad beans require early sowing so they can begin growing before the coldest part of winter. Nevertheless, I believe that in colder regions broad beans should be sown toward the end of February so they grow later, and Crecentino recommends not sowing them after December if the region is hot and dry.

Broad beans should be sown as follows. First, the land should be well plowed. In some regions, broad beans are then sown like wheat—that is, the seed is scattered—but this is not an efficient method. Broad beans are best sown in a furrow, in holes that are five or six fingers deep, approximately a foot apart, with a maximum distance of a foot and a half. A two-foot space should be left between furrows, providing sufficient space for walking when weeding, working the soil, and harvesting the broad beans without breaking them. It is preferable to sow them in a straight line, much like planting grapevines. Four or five broad bean seeds should be placed in each hole, leaving four or five finger widths of separation between each one to accommodate the many shoots they are likely to produce. It is said that broad beans sprout and grow better when sown together like brothers, and thus many people store broad beans in their pods and sow each pod intact in a hole of its own. I disagree with this practice, however, as seeds should be individually selected to obtain the best and largest, and more separation between seeds is generally needed. If sown within the pod, some broad beans will grow large, while others will remain small. Several agriculturalists contend that to grow good, large, tender broad beans in abundance, one should use the olive water technique prescribed for chickpeas. Since this is indeed beneficial, I recommend that broad beans be planted only after they have been soaked in the olive water long enough to become tender and begin to sprout. This allows them to grow faster in soil, even if they will not be irrigated.

Even though broad bean plants bloom a number of times and produce both early and late, the largest beans from the first harvest should be selected for seed. These are usually larger than beans from subsequent harvests, which tend to be stunted and damaged, and thus more suitable for eating than for seed. Once selected, broad beans should be grown under the same conditions indicated for chickpeas. They need not be watered unless they are planted in extremely dry places, but seeds sown in nonirrigated areas should have been grown under the same conditions. If broad beans are planted where they will be watered, seed grown in either irrigated or nonirrigated land may be used. Further, according to Crecentino, broad beans should be planted when the moon is full or waxing.

Only when the plants are four finger widths high should they be weeded and the soil surrounding them loosened. The soil should be loosened when the weather is dry so the plants are not adversely affected. The more the plants are weeded and the soil worked, the larger the fruit will be and the thinner the hull of the grains. Theophrastus and Pliny both maintain that rain damages almost all plants in bloom except broad beans, although rain is harmful to broad beans once they have finished blooming, as it kills the growth.

Broad beans should be harvested during the waning moon when they are completely dry. While still green, they should be cleaned and stored like chickpeas. Before being stored, however, they should be cooled off to deter grubs and other worms. If they have grown in fields to be sown with wheat, they should be flattened and plowed while in bloom so they can rot and fertilize the soil—a technique agriculturalists contend is excellent for enriching the soil. The same can be done with dry broad bean straw, though it is not as beneficial.

Regarding broad beans' effects on health, fresh broad beans are cold and cause heavy humors, headaches, flatulence, swelling, and bad breath. Dry broad beans cause the same problems and also compromise digestion and blood quality. Those who eat broad beans excessively, suffer from high fevers that cause frightening dreams. The water in which broad beans are boiled cleans and softens the body: topically, it eradicates ringworm, and if consumed it clears the lungs. Adding mint, cumin, or oregano when boiling broad beans eliminates the flatulence problem. Dry broad beans are fattening, especially in bread, while the fresh ones are not. In Lombardy, where broad bean bread is eaten extensively, there is a saying that someone who is overweight must have been eating broad bean bread. In fact, oxen are fattened with ground broad beans, or with broad bean flour combined with hay or straw, fifteen days before they are slaughtered.

Broad beans are also used routinely as remedies. It is said that a paste made from broad bean hulls applied to the lower part of an infant's body prevents hair from growing in that area or removes hair already grown. If broad beans are roasted, ground, and applied to the temples, they obstruct the flow of bodily fluids from the head to the eyes. If broad beans are cracked in half and placed on wounds, they stop bleeding, and if placed on a dog bite or that of any poisonous creature they both stop the bleeding and relieve the pain. Finally, broad bean flour applied topically is excellent for the chest, alleviating coughs when first stewed with garlic and, if used regularly, clearing the vocal chords.

LENTILS

Lentils should be sown in light, loose, dry soil, since moisture rots them, although they can be planted in rich soil if it is dry. They should not be planted in soils that produce many weeds, and they grow a greater number of grains in cold as opposed to warmer regions. Much like spring wheat, lentils should be sown in November in dry, hot regions and in February or March in cold, humid lands. Always they should be planted during the rising moon, some say on the twelfth day of the rising moon. Lentils sprout sooner and larger if placed in dry manure for four or five days before planting. They should be sown very lightly and preferably in straight rows, similar to the method indicated for broad beans. There is no need to weed or loosen the soil around lentils, however, because they grow tall very quickly and choke off all the surrounding weeds.

Lentils should be harvested in June to prevent grubs from causing damage, at which point the following method can be used to assure their preservation for seed. After the lentils are harvested and cleaned, they should be placed in a cauldron of water to separate those that are sterile, which float, from those that are fertile. Then the fertile seeds should be rinsed thoroughly with a mixture of asafetida roots, which Columella calls *"silfio,"* and vinegar; combined in a mortar; then dried in the sun. Once the lentils are completely dry and cool, they should be placed in earthen jars that have previously held oil, then sealed with plaster or clay. It is said that covering the lentils with ashes prior to sealing preserves them well so that when it comes time to sow them, they will be full and healthy. If there are huge quantities, the lentils can be stored in granaries. The best lentils are the larger ones, and the ones that absorb the most water when boiled. The wider, white lentils and those that do not turn dark when placed in water are also good.

Regarding their effects on health, lentils are cold and dry, provide substantial nourishment, but cause severe indigestion. They produce weak blood and are detrimental for people who suffer from epilepsy, abscesses, or illnesses caused by black bile, such as melancholy. Further, lentils can cause headaches, flatulence, and, if consumed in excess, frightening dreams. They can also cause shortsightedness, stomach inflammation, and constipation, particularly if boiled in rainwater. Those who have varicose veins in their legs or suffer from a form of a quartan fever also should avoid lentils.

In terms of their curative value, by attenuating both blood and heat and increasing melancholy, lentils mitigate impulsive tendencies to fly into a rage. For topical applications, lentils should be well cooked, or at least softened. When further moistened and spread on inflamed areas, they reduce the swelling; and a green lentil plant applied to a bleeding wound stops the hemorrhaging. Finally, lentil broth is beneficial for people who suffer from blood diseases, backaches, and other pain-related conditions.

LUPINES

Some varieties of lupines are wild and found in the mountains, while others are domestic. Both wild and domestic varieties are planted for one of two reasons: to fertilize wheat fields and vineyards or to be consumed. As a fertilizer, lupines excel, involve minimal cost and labor, and, unlike other fertilizers used on grapevines, do not ruin the taste of wine. To use lupines as a vineyard fertilizer, sow them during the grape harvest and plow them into the soil when they begin to sprout shoots—a recommendation more for vines to be trussed in trees or on structures than for those to be grown on the ground. When used for fertilizer, lupines should be heavily sown, but when used for seed or consumption they should be planted sparsely, because they produce numerous shoots. To fertilize wheat fields, lupines must be planted early and picked in bloom, around May; after the pods are harvested, the remaining straw can be plowed into the soil, thereby enriching the land, but not as much as when the lupines are green. To provide seed or food, the earlier that lupines are planted, the better. In general, they should be sown early and harvested late. According to Theophrastus, Columella, and other agriculturalists, lupines should be sown as soon as other crops are harvested. Because cold weather can adversely affect lupines, they should be sown earlier in cold regions so they will be established before the heavy rains and snows of winter arrive.

Lupines thrive in warm rather than cold climates. They prefer dry, loose, sandy soils to heavy, sticky, claylike or argil land. Theophrastus claims that lupines do not thrive in well-worked soil, because it causes them to grow excessive foliage. From the moment they are sown until they are harvested, there is no need to weed or otherwise care for them, as they choke weeds. They are so hearty that even if they are covered with a huge rock after sowing, they will sprout and grow well. It is good to plant lupines near roads and to enclose fields with them, since they are so bitter no livestock will eat them. But be aware—because they have a single root, if they are cut or touched by a farmer's hoe the entire plant will dry.

Lupines should not be harvested until it is humid, as otherwise the seed falling from the pod will be lost. And although they bloom three times, the first seed is the best. Even so, after each harvest the lupines should be dried on the threshing ground then stored in a very dry place to keep them from worms, preferably where there is smoke, which helps preserve them.

Regarding their nutritional value, lupines are hot and dry. When either ground or cooked, they provide excellent sustenance for oxen during the winter, and for people in lean years if no other grain is available. If they are to be consumed, they should be soaked in hot water, changing it repeatedly until they become sweet.

For medicinal purposes, the bitter varieties are better than the sweet ones. Specifically, bitter lupine flour kneaded with honey and applied topically to the stomach combats worms. Lupine paste or the water in which lupines have soaked can be applied to the face to remove blemishes, and people who eat them regularly have good complexions. Cooked in vinegar, they restore color to scarred skin. Drinking the water in which lupine roots have boiled stimulates urination and the passage of kidney stones. Moreover, lupines can be used to exterminate various insects and vermin. They can also cure scabies in animals if the lupines are cooked in olive water and spread on the affected spots. Also, water lupines have soaked in eradicates bedbugs and effectively exterminates ants when poured on infested areas.

VETCH

Vetch is best grown in cold rather than hot regions. It prefers light, dry soil, and thrives when sown in January or February, although in dry soils, it can be planted before winter. Theophrastus suggests that sowing it at the end of winter makes it better tasting, easier to digest, and more nourishing. According to both Palladius and Vicencio, vetch planted in March is harmful to livestock. It should be planted sparsely, as should all legumes, in previously cultivated fields. (In Italy there is a strain of legumes called purple [or blue] vetch, which I have not seen in Spain and do not specifically discuss in this book.)

Regarding its nutritional and medicinal properties, dried vetch is good for fattening livestock, especially cattle. Its moistened flour and grain provide good sustenance during the winter, particularly for oxen. Mixed with rye flour, vetch generates abundant milk for nursing livestock, although Aristotle cautions that it should not be given to impregnated animals because it increases labor pain and may endanger them when giving birth. Vetch creates flatulence if eaten whole, and if consumed on an empty stomach it can damage the spleen. Finally, applying its flour to wounds helps in healing, while kneading its flour with wine and spreading the mixture on pimples eliminates them.

MILLET (*MIJO*)

According to Crecentino, there are two varieties of millet. One variety, which takes three months to grow, is spring millet, or "*tremesino*" ("three months"); the other variety grows in about forty days. Both varieties favor cool rather than hot regions, and well-irrigated locations, although water can substitute for cool lands and cold temperatures can make up for a lack of water. Millet prospers in rich, loose soils and does not grow well in dry soils, particularly if they are heavy or sandy. It does particularly well in places that tend to be cloudy, such as along riverbanks or in valleys. Its fields must be well fertilized and worked, because millet can diminish the soil.

Spring millet should be sown at the end of February or during March, while the forty-day millet, which prefers more humidity, should be planted by mid-May. The soil should be very well plowed, as indicated for wheat, and sown sparingly, when conditions are humid rather than dry. The seed should be full, dry, well ripened, and golden or a bit white, rather than grayish or black, which indicates improper ripening. After it has sprouted, the millet should be irrigated once a week until the spikes appear whitish, indicating they are starting to ripen. If millet cannot be irrigated, to ripen it must be sown in a humid and cool region. Wherever it is sown, millet must be safeguarded from birds, especially pigeons, blackbirds, and sparrows. It should also be weeded frequently, since rich lands tend to produce many weeds.

Further, Crecentino asserts that this crop can be sown between broad-bean furrows. The combination is not appropriate, however, since beans do not need to be irrigated whereas millet does, at least here in Spain.

Both varieties of millet should be harvested before the spikes dry completely, because subsequently the grain becomes very smooth and falls off. After harvesting, millet should be taken to the threshing ground in sheaves, with the spikes facing the sun to aid in drying. The sheaves should not be stacked, since then they will become heated and the millet will be ruined. Some farmers reap only the spikes, cutting down the remaining stalks and allowing them to dry in the sun for four or five days, then storing them as straw that provides excellent sustenance for livestock. In either case, after the millet has been reaped, threshed, and cleaned, the grains must be left to dry completely in the sun, for if millet is stored without being fully dry it spoils easily. Varro claims that storing millet in silos can preserve it for a hundred years. Another good method is to store it in earthen jars well sealed with plaster and placed in a nonhumid location.

Regarding nutritional and medicinal uses, millet is cold and dry. Freshly baked millet bread is excellent nourishment and healthy for the stomach. It stimulates urination. Applying the warm grain to the stomach alleviates aches. Roasted, ground, and cooked in good meat broth, it makes a unique stew, especially if saffron and cinnamon are added. All who eat millet regularly grow fat, and thus it is excellent feed for pigeons and chickens. Magnino Milanes states that it is hard to digest, causes constipation, and engenders melancholy, and thus must not be eaten by people with epilepsy. For those who suffer from nerve pains, however, heated millet placed in a pouch and applied to the affected area is very helpful and also cures diarrhea.

BROOMCORN MILLET, OR PANIC GRASS (*PANIZO*)

Broomcorn millet is similar to millet in terms of attributes, and benefits from the same methods of sowing and cultivation. The best time to sow broomcorn is from February to mid-April, although it can be planted as late as May or June, when barley is ordinarily harvested, and in the same fields after the stubble is burned and the soil is plowed and fertilized. Because broomcorn requires irrigation, which depletes soil by carrying away nutrients, before sowing it is best to mix into the soil some well-ripened manure or mire from flowing springs or rivers and let the running water blend the fertilizer throughout the soil, enriching it. The broomcorn should then be planted very lightly, weeded frequently, and harvested and cleaned in the same manner as millet. It can also be sown among trees or irrigated grapevines, much like millet.

Although broomcorn's qualities are similar to those of millet, it not as good for sustenance. Even so, Diocles, a Greek physician living in the fifth century BC, called bread made from it the "darling of breads," meaning the most valued of all breads. (I imagine they do not have white wheat bread in his country.)

Broomcorn's nutritional and medicinal profiles are as follows. Bread made from it is fattening for people who consume it regularly and has a bad taste. Even when its grain is made into porridge with almond or goat milk, much like rice, it is fattening and causes constipation. It is best when cooked with meat fat, which softens it, adds flavor, and makes it less dry and less likely to cause constipation. Finally, ground broomcorn millet drunk in red wine, or boiled in goat milk and eaten twice a day before consuming other foods, is beneficial for diarrhea.

Borona, a legume found in the mountains of Biscay, has qualities very similar to those of broomcorn millet. What is discussed here also applies to its grain.

SPANISH PEAS

Spanish peas look much like chickpeas in color and size, except they are more angular. There are two or three types of Spanish peas, all of which favor heavy, moist, loose, and well-worked soils and grow best in rainy areas or humid places such as riverbanks. They are best sown in October and harvested at the end of a waning moon's cycle. Some spread along the ground, and for these, small poles should be used so the stalks climb like pumpkins rather than crawl along the ground and rot. If not completely dry when harvested, they should be left on the threshing ground until fully dry, then stored in a very dry place to prevent infestation by worms.

Regarding their nutritional and medicinal uses, Spanish peas provide good sustenance for oxen and sheep. Farm laborers often cook them in the same way as chickpeas then use them mixed with other grains to make a hearty bread. Pregnant women should avoid fresh Spanish peas, however, because they are harmful if eaten in excess.When eaten fresh, this legume can damage teeth and generate vitreous humors and bad breath. When dried, Spanish peas are good for cooking during Lent. But they can be stored only for a short while, because they tend to become wormy.

About Spices

ANISE

Anise thrives in warm or temperate climates. As Crecentino indicates, it is known as *"hinojo romano"* (roman fennel) in Italy, where it is sold by the bunch in the streets. It favors soils that are rich and somewhat moist. When grown in other soils, they must be fertilized and irrigated; but fertilizer, while fostering growth, compromises aroma and taste. The best type of fertilizer consists of ashes, river mire, or aged manure that has lost its powerful stench.

It is best, however, to plant anise lightly along riverbanks or in other fertile areas in February or March, using seed no older than two or three years. The soil should be well worked, with clods of earth completely broken up. The plants should be weeded when small, and safeguarded against birds, moles, and ants. Scarecrows can be used to frighten birds away, but moles remain a concern since they eat the root, and ants do their share of damage as well. Anise is harvested in May or June, before it is completely dry, after which it may fall off the stem and be lost. Once harvested, it should be cleaned in a clear, level area to prevent it from becoming mixed with dirt or pebbles, then left to dry. Anise can be preserved for four years, beyond which it spoils.

In terms of its nutritional and medicinal uses, fresh green anise—which is very sweet, tasty, and good—stimulates milk production in breast-feeding mothers. Eaten early in the morning, it eliminates bad breath and is very refreshing. Whether fresh or preserved, anise also deopilates the liver. Water in which it is boiled cleanses the stomach and adds flavor to any food; placed beneath a loaf of baking bread, it gets toasted, adding flavor to the bread. Pillows stuffed with anise straw provide a delightful scent and help eliminate nightmares, as does placing a small bundle of it at the head of the bed. Anise gets rid of chills and flatulence. If consumed before a meal, it generates a healthy appetite; eaten after a meal, it aids digestion and is effective against phlegm, vomiting, and burping. When burned as incense, it alleviates headaches, earaches, uterine pain, and "mother's aches." Consumed orally, it induces sleep, stimulates urination, and helps break up kidney and bladder stones. It is good for dropsy as well, since it

hydrates the body. Pythagoras claims that anise is so effective in treating epilepsy that merely holding it in one's hand prevents fainting spells and other related maladies. Washing oneself with anise, either ground or dissolved in water, thins the skin and the nostrils. Even if done occasionally, it improves jaundiced skin.

CUMIN

In general, cumin thrives in the same soils as anise but is more difficult to cultivate, sprouting only in patches. The best time to sow cumin is in February or March.

From a curative perspective, cumin taken internally aids digestion and clears the urinary tract. Drinking wine in which cumin, dried figs, and fennel have been boiled alleviates intestinal pain. Cumin cooked in wine and placed hot on the lower extremities relieves the symptoms of a urinary disease called *estranguria* in Spanish, which causes pain and difficulty with urination. There is also a type of rustic cumin that alleviates stomachaches if ground and drunk with water.

The best cumin, according to Pliny, is *carpentania,* a variety found in parts of Toledo. Although sown like the others, it should not be irrigated, because irrigation makes it less potent. When consumed, it alleviates stomach pain and effectively treats diarrhea. If drunk with water regularly, it cures jaundice. Pliny observes that it is used extensively by people who want to change their facial color. Finally, mixed with honey and applied externally to a child's stomach, it relieves indigestion.

CARAWAY

Caraway is planted at the same time as cumin and favors the same soils. This spice is a good antidote for flatulence, particularly when it has been ground with foods that produce gas, such as turnips or cabbage. It also soothes the head and assuages intestinal and stomach pain caused by cold weather.

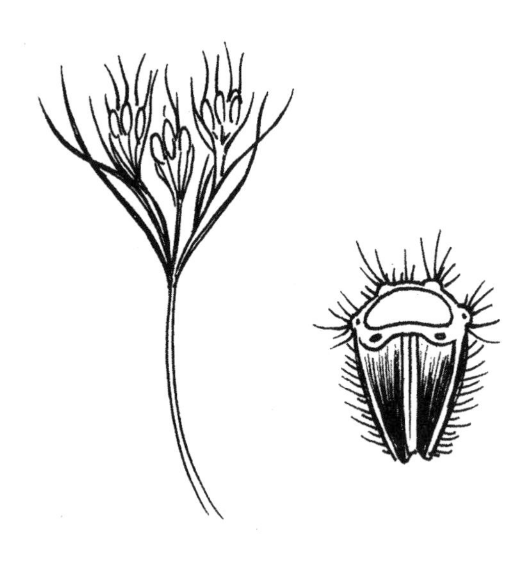

About Flax and Hemp

FLAX

Flax is beneficial for people in diverse ways, even though it is more damaging to soil than any other crop. Virgil emphatically asserts that regardless of where flax is sown it is detrimental to the land. Columella agrees, and adds that it should not be sown unless the land has been leveled. Two methods remedy the damage flax causes to land. The first method is to allow the land to lie fallow for a lengthy period of time. The second is to fertilize the soil extremely well.

There are two varieties of flax: a winter variety that is sown before the beginning of winter, widely known as *vayal* (meaning neither steeped nor soaked), and a variety planted in the spring, around February or March, known as *regantio,* because it is irrigated. Flax grows well in areas not suitable for other seeds, such as high elevations where neither grapes nor grains will grow. Both types flourish in well-plowed, leveled soils. Crecentino suggests digging as deeply as possible during the first plowing; less deeply during the second; and lightly during the third. *Vayal* flax requires little fertilizer and prefers fairly moist soils that are not very rich, such as sandlike mire or moist meadows. The richer the soil, the thicker and taller the flax, while the lighter and looser the soil, the thinner and more delicate it will be. This variety of flax prefers warm lands that do not freeze, as frost is detrimental, and it can also be sown around March in irrigated lands. *Regantio* flax should never be planted in dry areas. It thrives in rich, moist, irrigated soil well fertilized with ashes and aged manure, preferably from horses rather than other livestock as goat manure, for example, makes flax grow taller but too coarse for spinning.

Flax should generally be planted around March or April in colder regions—after the danger of frost has passed, since if the plant freezes before sprouting, it will sprout very unevenly. It can be sown on lands that grew turnips harvested for Lent; after the harvest, the soil should be plowed two or thee times before planting the flax. Flax seed should be fresh, yet because it is moist it can be very easily ruined. Ideally, the seed should be under a year of age though never older than two years. It should have been kept in a dry place and stored in a well-sealed earthen jar or other container keeping it from mice or other rodents. Flax should be sown very evenly and densely; the denser the planting is, the thinner the fiber and the cleaner the flax. Before sprouting, it should be safeguarded from birds through the use of scarecrows or anything that moves with the wind. Flax should be weeded when small, especially the irrigated variety, since as it grows, so do the weeds. When the plants are tall, special care should be taken not to trample them because each one that falls will most likely be ruined.

After the sowing, it is sufficient to irrigate flax once a week, either in the morning or evening, until it begins to manifest some color, then stop irrigating. Although some people harvest flax when it is still green, this practice should not be followed, since neither the flax nor the seed is ripe so early on and the seed is likely to either quickly

spoil or never grow when planted. Flax should be harvested before the husk is naturally opened, at which time the seed too easily falls out. According to Pliny, flax is ripe when it is yellow or when the seed is plump. When harvested, the flax should be gathered into small bundles and positioned with the grain downward, a practice that fattens the grain. The bundles should be protected from moisture, since moisture rots the grain, particularly if the bundle gets wet inside, leaving much tow and very little flax with short fibers. If flax gets wet, spread it out to dry then gather it again, removing it as soon as possible to keep mice from eating it.

If the flax seed is left on the stalk, it will become heated and damaged. Therefore, some shake it out on sheets, which is the most efficient method—unlike in Arroyomolinos, where a wooden comb is used with teeth so narrow it cannot properly separate the seed from the fiber, resulting in much lost seed. After removing the seed, it should be soaked preferably in still water, with heavy weights placed on it because it is very light. If soaked in still water, the flax will be ready in four or five days, whereas if it is soaked in running water it will take two additional days. When soaking the flax, ensure that the weather is clear and sunny, because sudden rains can carry the flax away in the arroyos. It should also be noted that the water in which flax is soaking should not be consumed, because it is noxious. To dry the flax, gather it into small bundles, tie them at the top, and place them outdoors in the sun, with the heavy grain angled downward. After the flax has been well dried in the sun, it should be stored in a humid place then pounded so the fiber stretches. If pounded when completely dry, on the other hand, the fiber will tear and much will be wasted. While pounding, remove and discard the roots.

Regarding its medicinal and other uses, flax is hot and humid, and relieves a cough if boiled in water and eaten with honey, although it causes flatulence. If spread on abscesses and lumps, it reduces the inflammation. Flax is also used for making clothes, and after they have become worn, it can be cleaned and shredded for use in dressing wounds. When cut into pieces, cleaned, and bleached white, it can be used to make paper as well, another item indispensable for human existence.

HEMP

Hemp is very similar to the irrigated variety of flax, preferring comparable soil and climate conditions. To grow hemp for use in rope, halters, or cables, the soil should be rich, loose, fertilized, and moist, or in an area that can be irrigated, and the temperature cold rather than warm. If sown in rich soil, hemp grows thick and tall, with good strong fibers for use in clothing. It should be planted densely, so it will grow tall and thin, with fine fiber; if planted sparsely, it will have too many stalks and be of a lower quality. In cold regions, hemp should be sown around March, while in very cold regions it should be sown in mid-April and in warmer lands around February; it will rarely fare well if planted before December. The crop should be irrigated like flax and weeded when small, after which it grows so tall that it chokes off the surrounding weeds. It should be harvested when the seed is ripe then cured in the sun and prepared like flax, although hemp is coarser and harder to wash because it is more easily torn.

Regarding its medicinal properties and other uses, hemp seeds are hot and dry, and people who eat them often lose reproductive potency, as Avicena and Pliny reveal. Hemp also produces shortsightedness and is injurious to the stomach, yet it is commonly known to relieve coughing. In addition, it provides sustenance for small birds raised in cages and good fiber for making cords, ropes, halters, and heavy cloth used for sacks.

CHAPTER TWO

PLANTING AND HARVESTING VINEYARDS

Favorable conditions for growing, gathering, storing, and preparing grapes

THERE ARE FOUR basic types of grapevines. One type is trussed in trees and thrives only in humid, very dense regions such as Lombardy, other parts of Italy, or similar areas, while another is trussed on clinging vines or structures. Both types thrive under the same conditions and methods of cultivation. A third type of grapevine, found in the Campos region of Castile, grows on the ground, making it the least desirable of them all. Wine derived from these grapes retains the flavor of the earth, and the grapes are easily damaged, spoiling with only minimal moisture. This type should only be sown in very loose, sandy soil, allowing rainwater to seep into the ground rather than remain on the surface and spoil the grapes, or run off too quickly. Under proper conditions, this type of grapevine produces much fruit. The fourth type is composed of more common varieties that grow upright without support, much like small trees. These vines grow neither very tall (in contrast to the varieties trussed in trees or on clinging vines or other supports) nor very short (as do those on the ground). They thrive in temperate land that is neither very humid, which is best for clinging vines, nor very dry, which is preferable for vines that grow on the ground. The vines that grow upright, which are the easiest to work with, thrive in both higher and lower elevations, on hills and plains, and they produce good wine. Trussed vines thrive in plains, while those growing along the ground favor the highlands.

For all types, the quality of the wine is determined by the soil, as much as if not more than the grapevine itself. In fact, a bad grapevine planted in rich soil produces better wine than a good grapevine planted in inferior soil. To ensure that they produce the best wine, good grapevines should be planted in the best soil possible in warm rather than cool regions.

Guidelines for planting and harvesting all four types of grapevines follow. I will provide recommendations as well for enriching the soil, storing the unripened grapes, making raisins, fermenting wine, and assisting as wine turns to vinegar. Included are both culinary and remedial properties of grapevines.

Varieties of Grapevines

The varieties of grapevines are extensive, and each country with a temperate climate has its own. Moreover, similar grapevines are known by different names in various regions and their names change with the

generations. I believe it is therefore best to concentrate on the individual qualities of major varieties, because these have remained constant throughout time. Descriptions of the main varieties that follow are organized by the three basic colors of grapes: white, dark, and red.

VINES YIELDING WHITE GRAPES

The best of all vines yielding white grapes are ALVILLAS, known for their round, small fruit that grows in very compact bunches with spots ranging in color from dark to tawny. These vines grow best on plains rather than in highlands, and because the grape is dry and firm and its stems stiff and brittle, just a little wind can break many branches. The vines grow tall, and therefore the grapes rarely spoil. If grown in groves and cool areas, they are preserved on the vine longer than any other variety. Ideally, though, these grapes are grown in humid regions, because no matter how much it rains they rarely spoil or burst, provided they are planted at high altitudes where water quickly runs off the leaves and the vines dry well when exposed to the sun. In areas conducive to fast ripening, this variety of grape ripens before any other, and where fruit ripens late it can be preserved longer. The *alvilla* has several additional advantages over other grapevines: it grows better in dry, sandy, gravelly places; it carries grapes and develops branches well; it is perennial; it grows well when trussed; and when pruned after the harvest, it produces small bunches of grapes that are good for sour grape juice. Wine made from these grapes is very clear and fragrant, with exquisite color, taste, and longevity. The wine can be further improved by mixing in *moscatel, cigüente,* or other similar varieties.

TORRONTÉS have small seeds and are more transparent than other grapes. They grow in small bunches that are not very compact and flourish in high, dry places relatively free of wind, because their skin is very thin. Their stems are so fragile that the majority of *torrontés* fall from the vine and at harvest time must be gathered from the ground. Although the grapes themselves are not very tasty, the wine made from them is better than any other made from white grapes—clear, fragrant, smooth, and able to be preserved a long time.

The MOSCATEL lineage of grapes is characterized by its musky flavor and odor. They grow in compact bunches with very tender seeds. If the soil is extremely rich, they have a tendency to spoil; hence, they are best suited for sandy, dry highlands, provided these are not too elevated, because no grapevines flourish at very high altitudes. Wine made exclusively from *moscatel* grapes is excessively rich and sweet, but mixed with wine made from other types of grapes it is good and fragrant, improves with age, and can be preserved a long time. In addition, any grapevine variety grafted with *moscatel* takes on some of its fragrance. Since these grapes are very tasty, vineyards that grow them should be guarded as closely as possible.

The CIGÜENTE grapevine is very similar to the *alvilla*, in both appearance and attributes. These grapes, however, prefer hotter and drier lands that are not very rich. And the grapes are prone to spoilage, since their skin is neither as firm nor tough as that of the *alvilla*. Vines that remain unpruned carry more fruit on their branches than those that are pruned. Wine made from these grapes is very fragrant, clear, and keeps for a long time.

The JAÉN variety of grapevine produces large, very compact bunches of grapes with big seeds. Their skin is quite tender and often cracks and spoils with exposure to water. Thus, these grapes are best suited to dry, hot, sandy, and airy lands where there is little dew or rain, or, in the event of a downpour, the water is quickly absorbed. To ripen well, they should be planted where the sun shines intensely. When grafted with *moscatel* vines, they bear plentiful fruit and take on some of its fragrance. Due to the fragility of their skin, however, they should be harvested before heavy rains set in. Wine made from these grapes has good flavor but, in my opinion, cannot be preserved for more than a year.

Grapes of the HEBÉN lineage of vines grow in long, sparse bunches, have large seeds, and are more villous than any other variety. These grapes taste somewhat sweet and, when eaten, their fragrance permeates the nostrils. The fact that their bunches have few grapes is due to their late flowering, or "blossoming," period, after which much fruit is lost to the ravages of wind or rain. Hence, these grapevines flourish in warm regions where it seldom rains, provided the soil is dense and humid, although they have an advantage over other varieties in that they also do well in dry areas. They are best suited to areas that are covered and protected from winds, and also, since this variety rarely spoils, they tolerate lower altitudes. They produce a moderate amount of fruit and are perennial everywhere, provided they do not become too thin. This variety grows branches well and is good for trussing if planted in dense soil. Wine produced solely from these grapes is exceptionally sweet, indicating that *hebén* wine, like *moscatel*, should be combined with another type of wine, which it will infuse with a delightful fragrance. Its wine has a golden hue and substantial longevity, particularly if it is combined with wine from other superior grape varieties such as *alvilla, cigüente, torrontés,* or *castellano*.

VINOSO grapevines produce grapes that are large, sparse, and similar in appearance to the *hebén* variety. *Vinoso* grapes, however, are very tender and, prone to spoiling, thrive in drier regions. Large quantities of wine are produced from them—hence, their name, *vinoso,* meaning full of wine. The wine is very clear, smooth, of moderate longevity, but uninteresting. If combined with wine made from *alvilla, cigüente,* or *torrontés* grapes, it develops a wonderful color and taste, robustness, and liveliness.

CASTELLANO BLANCO grapevines are known for their small, compact bunches of grapes that are round and solid, yet tender. They thrive in dense, sandy, and gravelly soils. These grapes produce a good, though not excellent, quality wine.

MALVASIA, also known as "*más vale*" ("better"), grapevines grow small, compact bunches of fruit. The grape is round, dense, and fairly large if planted in good soil. Because it is tender and spoils easily, this variety produces a better wine if grown in dry lands.

The variety of grapevines called LAIRÉNES could, in my opinion, be better named *datileñas* (datelike), because the grapes they bear grow in bunches similar to dates. Although they yield more fruit in humid regions with rich soils, they spoil quickly in this climate if they are not maintained on high trusses. They flourish in dense soils too, provided they are moderately moist. If trussed, the leaves should be removed, permitting the sun to dry the fruit. While I have neither seen nor tasted wine made from this variety, I assume it would lack robustness and longevity. In my experience, this variety has few seeds and is best for producing excellent raisins.

VINES YIELDING DARK GRAPES

Although many varieties of grapevines produce dark-colored grapes, the best known are the *castellanas, palominas, aragonés, tortozóns*, and *herríals*. The CASTELLANAS yield grapes that are very dark, with typically small, compact bunches, although if the land is very fertile, they can grow larger bunches. The grapes have a small seed and very tender skin. The vines, which rest near the ground, thrive in loose, sandy, dry soils at high altitudes. They bear more fruit when the soil is rich, but then they are more likely to spoil. When planted in rich soil that is not excessively moist, the grapevines will produce a quality wine and will not spoil. This variety ripens more rapidly than any other bearing dark grapes. During exceptionally humid years generating excessive growth, wine derived from these grapes is superior to any other dark wine. It mixes well with other excellent wines, such as those made from *alvilla, cigüente,* or *hebén* grapes, resulting in a superb clear, smooth, and fragrant wine of significant longevity.

PALOMINAS grapevines produce black fruit in long, sparse bunches similar to those of the white *hebén* variety, which explains why in some regions this grapevine is called "dark *hebén*." Although best suited to the same type of soil as the *hebén*, it ripens later. The sun can severely harm *palomina* grapes; hence, those covered by leaves tend to ripen better than those exposed fully to the sun, which instead acquire a reddish color and remain hard. This variety does best when its branches are long so leaves do not have to be removed. The wine made from its grapes is very clear and particularly good in early summer. This variety generally does not preserve well because its grapes absorb the easterly wind. Nor does its wine combine well with those of other varieties of grapevines.

ARAGONÉS grapevines yield dark, dense grapes in large, very compact bunches. If grown in flat, rich lands, the vines yield plentiful fruit, but if grown at high altitudes in sandy, dry soil they may not generate much fruit, although, provided the soil is substantial, the grapes will make a clear, smooth wine. By contrast, the *aragonés* grapes grown at lower altitudes produce a dark-hued robust wine that can be preserved for long periods of time but tends to cause kidney and bladder stones. It is better when mixed with white wine.

The TORTOZÓN and HERRÍAL grapevine varieties yield abundant and large bunches of fruit with extremely large seeds, similar to the *aragonés* variety. To produce the best possible wine from them, one should plant the vines on hills in sandy, gravelly, and dry areas. The wine made from these grapes is basically a pantry wine that cannot be preserved for long.

VINES YIELDING RED GRAPES

ALARIJES are grapevines with tall stalks, much like the *alvillas*, and they flourish in the same type of soil. The grapes are bright red, and because bees frequently consume them, they are sometimes planted near

beehives so that in lieu of flowers bees can resort to eating them for nourishment if necessary. They produce a very red wine, which preserves well and rarely sours, but has many flaws associated with its partiality to the easterly wind. For example, it is murky and can turn to vinegar.

Soil Conditions and Planting Techniques

Soil for grapevines should be sweet, with good flavor, and a source of water that is neither bitter nor salty, since the wine's flavor will reflect these properties. Soil where trees and brambles flourish, either cultivated or wild, is also suitable for vineyards. Soil that splits and cracks in the heat of summer, however, is detrimental to grapevines, because the sun and heat permeating the cracks can desiccate and burn the roots, except in areas where irrigation is feasible.

Irrigable areas with nutrient-rich soil, propitious for producing much fruit, are

suitable for grapevines. Soil that is dense while also loose and light is likewise suitable even if the topsoil appears poor, provided it is dense and substantial below, because the soil underneath sustains the plant while the topsoil protects it from excessive cold and heat. Although grapevines rarely take root in clay, topsoil that is sandy or loose with clay underneath is beneficial for grapevines when well-ripened manure and much soil are put in the planting holes. If the soil is shallow, the vines should be planted where both soil types meet, and should not be trimmed. The following year, when they have grown two or three sections long, they should be buried in the clay and covered again with soil.

Gravelly and rocky surfaces are detrimental to grapevines, because rocks by nature absorb heat in summer and cold in winter. Thus, if there are many rocks, it is best to pile them up away from the vines or dispose of them outside the vineyard. Some pebbles or gravel deep in the ground, however, is very beneficial for grapevine roots, keeping them cool in summer and expelling excess water in winter. For this reason, farmers knowledgeable in the art of planting place three or four rocks around the roots, regardless of the type of vine.

Sandy soils yield little fruit, but they produce grapes that make excellent wine. The best grapes for this soil type are those that commonly spoil with moisture, such as the *jaén* or similar varieties with tender seeds. Sandy soil with sweet water underneath is very good for growing grapes that produce excellent wine. Lands with argil or red, hard clay are deleterious because they retain excessive moisture in winter even if seldom watered, and become very hard and dry in summer.

The best soil for grapevines absorbs water quickly during irrigation or a rainstorm, and retains a moderate level of moisture. Moreover, farmers concur that old roots must be removed from formerly dense vineyards (called *herriales*) so these do not obstruct new plants. When planting grapevines, both good soil and high-quality plants ensure that wine produced from them will always be superior and the farmer will not be scorned.

LOCATIONS FOR VINEYARDS

There are two location options for planting vineyards: in the highlands or on the plains. Vineyards planted on plains yield more fruit than those at higher elevations. However, the wine from grapevines planted in highlands is far superior, more fragrant, and has greater longevity. In the mountains, vineyards exposed to the north wind produce more fruit than those exposed to the noonday sun because the north wind keeps the grapes cooler. Vineyards exposed to the noonday sun produce superior wine because they are in a better location for ripening—consequently, the grapes remain drier and thus less prone to spoilage. Valleys, especially deep ones, are the least desirable sites for vineyards because there grapes often spoil and remain greenish, without proper color or flavor. So although grapevines in valleys produce much fruit, the smaller quantities grown in highlands are far more valuable, unless the valleys are exposed to the sun. Nonetheless, since there is generally not sufficient land to be selective about vineyard sites, we shall discuss the types of grapevines best suited to various locations.

The best grapevines to plant on the plains and in valleys and humid regions are those that grow long stocks, with bunches that are not very compact and firm grapes with hard, thick, and dry skin, such as the *alvillas, palominas,* and similar varieties. The best grapevines to plant in the highlands are ones with shorter stocks, tender seeds, and large, compact bunches, such as the *jaén, moscatel*, and *torrontés* varieties. Vineyards in hot regions should be exposed to the north wind, which provides cool air. In colder, humid areas, vineyards should be exposed to the noonday sun, providing more constant sunshine. Grapevines that can be damaged by dew and fog, such as the *jaén* variety, should be exposed to the east; conversely, varieties like *alvillas* and all grapes that are firm, dry, and benefit from fog and dew should face west. Grapevines such as the *alvilla* and similar varieties with hard stems that break easily should never be planted in windy regions, since they are easily damaged. Virgil maintains that westward-facing vineyards are never good. This counsel is not applicable to all grapevine varieties, however, for while it is harmful for some, it can be beneficial for others. Many farmers recommend exposing vineyards to the west provided the varieties of grapes are dry, firm, and benefit from dew and fog, such as the *alvillas* and *palominas*.

GRAPEVINE SELECTION

When planting a vineyard that may exist for fifty or even hundreds of years, the best plants and varieties should always be used, because the loss incurred by planting inferior plants far exceeds that of planting poor seeds of wheat, barley, or other such crops that are not perennial. One solution for improving plants is grafting, but grafting requires even greater effort and is sometimes more harmful than beneficial. Thus, depending on preexisting conditions *jaén* grapevines should be planted instead of *tortozóns,* or *alvilla* grapevines instead of *jaénes*. Furthermore, within each variety the best plants should always be selected. It is far better to persist in seeking good plants initially than to have to remove or improve those unwisely selected. Also it is best to plant three or four grapevine varieties in a vineyard so that if a particular variety fails to produce fruit, as often occurs in certain years, the vineyard will not be entirely unproductive. The different varieties, however, should be of similar quality, since a mixture of incompatible varieties produces inferior-tasting wines with little longevity.

Further, each grapevine variety should be planted separately from the others, creating two to four sections, because at harvest time it is arduous work to separate interspersed varieties. Due to time constraints, all grapes are usually harvested simultaneously, regardless of the fact that some may be ripe and even overripe, while others are green and even unripe, since grapes do not all ripen at the same time. In addition to being visually pleasing due to its organization, and having the aforementioned benefits, maintaining each variety separately facilitates pruning at the proper time. Most importantly, if each grapevine variety is separated, it can be planted in the soil that best suits it, which is impossible if they are all intermixed.

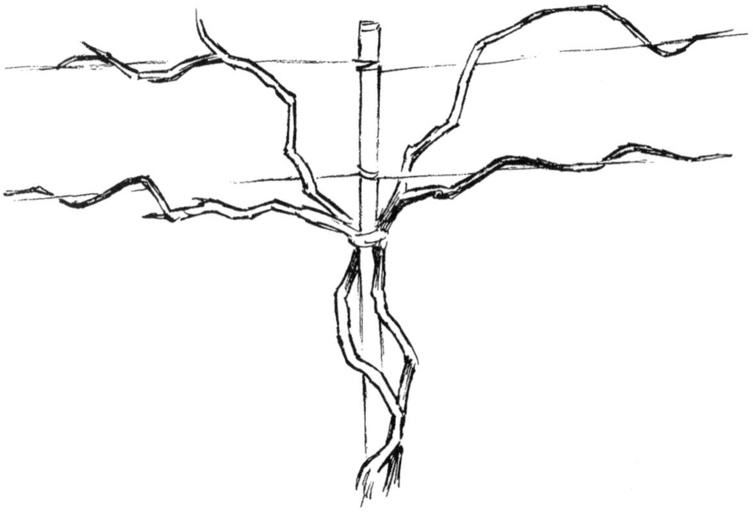

Basic Planting Methods

After selecting land for the vineyard, the farmer must prepare it, which may involve clearing the ground and removing grass, plant, and tree roots to ensure that these do not obstruct the tender new grapevine roots. Then the vineyard can be planted using one of three basic methods.

THE METHODS

One method for planting vineyards calls for the use of seeds, but according to Theophrastus, this approach leads to vines that are sterile, unruly, mature late, and must be grafted to be fruitful. Nonetheless, if planting seeds is the preferred method, either because seeds can be more easily transported over long distances than vine shoots, or for any other reason, the seeds should first be sown in an *era* (seed plot), because they are essentially as hard as wood. When after two or three years they have produced viable shoots, these should be grafted with the shoots of other grapevines to mitigate the deficiencies inherent in being grown directly from seeds. There are two types of shoots: *cabezudos* (large-headed), or pruned vine shoots, and seedlings with roots. Of the *cabezudos*, those that take root are far superior but not as reliable for transplanting as the seedlings, since many become lost whereas few seedlings fail.

Ancient farmers, aware of these outcomes, arrived at a third method for planting vineyards: they used the *cabezudos* to produce seedlings, which they called *maleoles* (malleolus), synthesizing the positive attributes of the *cabezudos* with the reliability of seedlings. The seedlings were grown in an arable plot that had been prepared between the rows of vines and referred to as a seedbed, or *almanta*. This is the same method used when leeks or cabbages are sown to be transplanted later in vegetable gardens, though in vineyards it is implemented as follows. First, the vine shoots are selected; it is ideal if the shoots have produced seedlings naturally, since these are the best for planting. Selection involves flagging the most productive grapevines when they first begin yielding fruit and continuing the practice over three years to factor out weather and other variables in selecting the most generative vines. On the marked grapevines, also flag the vine shoots or branches that produce the most fruit. Ideally, these will be vine shoots that grow at least in pairs, with very full bunches of large, good, dense grapes. For flagging, use a tie of some sort or, as Columella advises, a little vermilion or red clay diluted in enough vinegar to prevent the color from fading in the sun and rain.

To minimize site and climate differences before planting, carefully observe the vineyard's climate and site criteria then ensure that the vines for transplanting were originally grown under similar conditions. For example, for a vineyard exposed to the noonday sun transplant vines that were previously exposed to the noonday sun. If the vineyard is at a high elevation, plan on transplanting vines originally grown at a corresponding altitude. Similarly, if the vineyard is situated in a humid region, select vines accustomed to humidity. Ultimately, newly transplanted vines should also face the direction in which they initially sprouted, since then they will take root better and bear more fruit.

If it is not possible to select vines of similar site and climate orientations, use those grown in plains, provided they are not too tough, because plains-grown vines have the most strength and are typically of superior quality. The vine shoot should have plump, thick buds and short sections. It should be round, very green, smooth, moderately thin, and on a vine that is at least a year old so it will root well. In addition, the plant should be dense and contain no dry segments. Ultimately, the plant should be taken from the top of the grapevine and have no more than a two-finger length of growth from the previous year, because this is where the roots will develop.

If vine shoots, regardless of their place of origin, must be transported over a long distance, they should be well covered with a wet cloth to protect against the drying effects of sun and wind, particularly the easterly wind, which causes the most harm. Ensure too that the vine shoots do not rub against one another, which will harm the tips. If among the selected vine shoots there is a particularly valuable one that requires protection *before* being cut, a small basket filled with willow twigs or other tender branches should be placed around the vine by threading it through a hole from bottom to top. The chosen plant, once situated amidst the twigs, should be supported with additional soil and kept as is for a few days until it sprouts—at which point it can be cut from below, carried in its basket, and transplanted.

The most auspicious time to transplant is during the waxing moon, when the vines will take better root. I maintain that plants cut in the afternoon during the waxing moon have the highest survival rates of all.

And while some agriculturalists contend that vine shoots cut during the waning moon decay less rapidly than those pruned during the waxing period, they fail to make a strong argument. Actually, cutting has two objectives—to preserve the plant and to transplant it—each of which is optimally undertaken at a different time. Moreover, cut vine shoots should be transplanted from January through spring; prior to January, they lack sufficient maturity and strength. Young plants that are robust and have excellent roots, on the other hand, can be transplanted beginning in the fall and, if the soil is dry and hot, up until January. When pulled in the afternoon during the early phase of the waxing moon, they produce excellent vines that grow properly.

FROM SEED PLOT TO VINEYARD

Since planting seeds directly in a vineyard is not effective because vines produce little fruit if the soil below has not been properly plowed, to encourage root development it is best to grow seeds first in a "seed nursery" such as an *era,* or seed plot, then transplant them to the vineyard. The location and soil of the seed plot should be similar to that of the new vineyard. That is, if the vineyard is on a hill, the seed plot should be placed on a hill; and if it is on a plain, the seed plot should also be located on a plain; if the vineyard is on dry land, the seed plot should be located on dry land; if the vineyard is on humid land, the seed plot should be in a humid area. As a result, from the time they sprout, the plants will be accustomed to conditions similar to those where they will be transplanted. Further, if there is sufficient space, the seed plot should be located in the same area as the vineyard to ensure minimal variation in soil type and time for transplanting, ensuring that the tender roots are not damaged by the wind or sun.

The following is the best transplanting method. Dig a row, much like a ditch, knee-deep if the land is temperate, deeper if it is dry, and less deep if it is moist, because too much moisture from underneath spoils the roots and dries the plants. After carefully selecting the vine shoots according to the methods indicated, plant them deep enough so that at least five tip-lengths lie beneath the surface. They should be planted in a hole the size of a palm's width, because this allows the roots to grow very well, and the tips should be pruned to stimulate root growth. Many people twist the vine shoot based on the mistaken perception that this stimulates its growth, but it stresses the shoot and its lower tips, which should remain healthy for root growth. Others crush the heads a little when planting, but I believe this is just as detrimental. Any technique that requires killing part of the shoot inhibits rather than stimulates root growth. Vine shoots should be planted one foot or more apart in the seed plot, to prevent them from touching, which dries the plants. A few rotten grape skins or some barley grains should be combined with the soil to stimulate root growth. When planting vine shoots, a minimum of two or three tips should remain above the surface, so that if one falters, the others will grow.

Another planting method that requires less work but is not as effective or reliable is the following. Drive a sharp, hard wood or steel stake as thick as an arm into the ground, remove it, place the vine shoot in the hole, and then fill the hole with dirt and water. Crecentino maintains that it is best to use a hollow steel stake to make the hole rather than a solid one, to permit the removal of dirt without its becoming compacted. This method cannot, however, be used successfully in rocky soil.

While in the seed plot, the soil around the vine shoot roots should be repeatedly dug or plowed to stimulate growth. Further, while some people allow the vines just to take root before transplanting them, I advise letting them sprout as well. Each one should then be pruned of all sprouts to create a single vine shoot for transplanting in the vineyard.

When the vine shoots are ready to transplant, the following methods should be used. If the vineyard is located in an elevated or dry area, the holes should be very deep and not too far apart so moisture is better maintained. But if the vineyard is located on a plain or in a humid area, the holes should be dug farther apart and not as deep, permitting the sun to better penetrate between the grapevines. If the soil is moist, the holes should measure only four palms in depth, or a *vara* (about thirty-three inches), but if the soil is dry, the holes should be one-third deeper. If the vineyard is planted on a hill, the holes should be dug as deeply as possible, because rains will continually carry the soil to lower ground, meaning that if it is not planted at a sufficient depth the roots become exposed. Regardless of the location, the holes should be dug very wide, because the roots grow as large as the holes in which they are planted. According to Theophrastus, grapevines have slim, thin roots that rarely grow beyond the area that has been dug. Moreover, most farmers agree that it is preferable to dig the holes a year prior to planting, allowing the soil to absorb water and sunlight. Having been prepared in this way, the vineyard can be quickly planted, since the

seedlings simply have to be placed in the holes, which should be dug again and, following the planting, covered with soil.

Each seedling should be planted individually, ensuring that it is placed at approximately the same depth as it was in the seed plot. Then the entire old portion, and even some of the new, should be covered. Grapevines planted rapidly in this manner produce healthy plants, with growth as certain as those that have had their tops clipped.

Another method that may be used entails planting the seedlings in their permanent location from the outset. They should be chosen using the same technique described previously, but they need to be planted more meticulously to ensure success. Two plants of the same variety should be placed in each hole without touching, so that if one dies the other might still survive. If both take root, one can be removed and transplanted elsewhere. The holes should be the depth indicated above, according to the altitude of the land.

If the land is cold, add a little fertilizer combined with soil, but the manure should

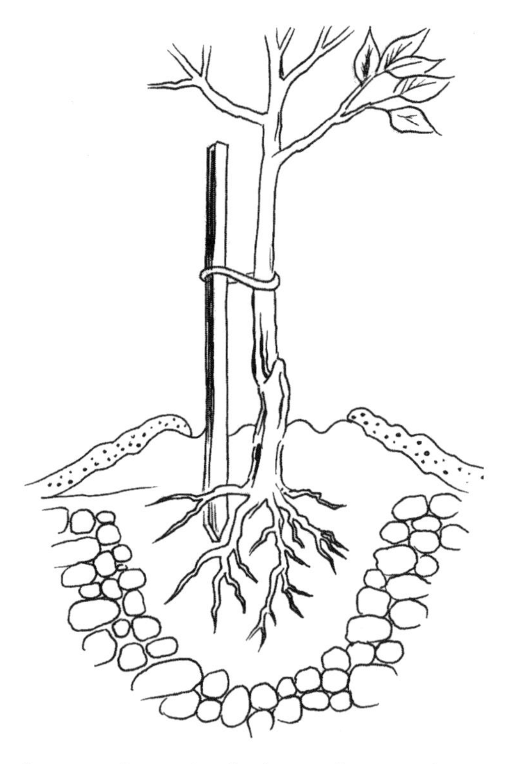

be very ripe so its freshness does not burn the vine shoots. If the soil is very rich, blend in sand and loose dirt at the base of the vine shoot. If the soil is loose or sandy, combine it with richer soil or white or red clay, using the same method. This procedure stimulates root growth and helps plants establish themselves. At planting time, four or five rocks, weighing about five pounds each, should be placed around the bases of roots but not too close together or they will obstruct root growth. Further, I recommend adding small round, smooth rocks or pebbles when transplanting seedlings, to help keep the roots cool and preserve moisture. After the vine shoot heads have curved, they should be planted at a palm's length or a little more, recognizing that curvature is better on the tips than on one of the sections branching out from the plant's internode. Ensure that the seedlings are placed in the center of the holes, permitting equal root growth in all directions. They should be planted vertically from the tip's curve, rather than horizontally or bent, because a straight grapevine has a greater chance of flourishing. Moreover, when plowing or hoeing, the vines planted vertically will not be as damaged as those that are sprawling. When adding soil, it is best to compact it around the lower layer, thereby securing the seedlings in the ground.

Grapevines should be transplanted using the same technique applied to trees. If the plants are placed in well-compacted dirt, the roots grow more quickly, the wind cannot move them around, and they are more certain to thrive, as Seneca sets forth in his letter beginning "*In ipsa Scipionis Affricani villa.*" If the planting is to occur in a somewhat dry and hot area, the holes should not be covered completely, thereby allowing the vines to collect and retain water until it is absorbed. If the land is humid and cold, however, the holes should be covered and level with the ground that surrounds them. As for small plants, they should never be covered, since they need to come to tolerate both cold and heat. Also, when seedlings are covered they sprout roots on the surface rather than underneath, and grapevines with surface roots dry up with little heat, are severely damaged by ice, and are easily uprooted by a plow. In fact, it is best to dig up new grapevines annually, in order to remove all surface roots, young and old. Vines that are not transplanted should be weeded and hoed underneath, where vine shoots have sprouted, allowing them to grow new branches. Instead of hoeing underneath the vines, some people cut the tops, removing the tips of the vine shoots, which is a more reliable technique. Others hoe and graft the vines, and for those that take root this is unquestionably the best method.

In dry or hot regions, grapevines should be planted in autumn, especially when they are transplanted seedlings. Vine shoots to be planted at that time should be cut when they first sprout from the leaf, since that is when they are especially robust. In cold and humid regions, transplanting should be done in the spring, or at least after January, when the vine shoots are particularly healthy. To avoid damage, they should be protected from sun, cold, and wind, particularly the newly cut areas. They too should be planted when there is no wind, especially no easterly or northern wind, and it is not very cold or sunny. If possible, it should be warm, calm, and cloudy but not rainy, or at least not rainy enough to cause the soil to turn to clay. It is most beneficial to plant between the first and the tenth day of the waxing moon and to daub the tips with ox manure to make the roots grow stronger.

To do well, transplanted grapevines also require a proper site in the vineyard. One must be able to dig around the roots so they can easily extend underground, for example. Therefore, the first soil layer should come from the ridge between the furrows, if it is not too dry, and it should be solidly compacted around the roots. If the vines are *alvillas* or a similar dry variety, they can be planted closer together. In dry regions, it is preferable to plant the grapevines in the space between the furrows, so they are not on the ridge; this will better protect them from the sun so they retain more moisture. In humid places, vines should be planted on the ridge, so they are thoroughly exposed to sunlight to dry excess moisture. In soil that requires plowing, the furrows should be straight, even, and wide so that neither the plow nor animals damage plants. So they will grow tall, all the small shoots that sprout between the leaves should be pruned, as should all the small vine shoots that some call "grandchildren" ("*nietos*"). Above all, this should be done with the vine shoots that will themselves be transplanted elsewhere. Finally, it is imperative not to plant young grapevines near laurels, hazelnut trees, or cabbage, because these can damage the vines so severely that new vine shoots will perish and branches of more mature grapevines will grow in the opposite direction. The attributes of grapevines are so contrary to those of cabbage that if someone has drunk too much wine, eating raw cabbage leaves eliminates the inebriation.

Recommendations for Cultivating Grapevines

When it comes time to cultivate newly transplanted grapevines, each type requires special attention at the outset so the plants can get the best start possible. A number of strategies can then be implemented universally to maximize the production of fruit. These include suitable methods for supporting the vines, shaping them, pruning and grafting them, and adding medicines and fragrances. To insure against loss, the farmer will also want to tend properly to curing vine afflictions, plowing and excavating, fertilizing, branch removal, and overall vineyard management. For a bounteous harvest, there is much to be done day by day throughout the growing season.

WORKING WITH GRAPEVINES TRUSSED IN TREES

Grapevines trussed in trees thrive in humid locations such as in valleys, along riverbanks, and in rich soils. At high altitudes, it is best if the trussed vines are of a firm variety, such as *alvilla,* since vines are prone to spoiling if exposed to too much rain. The *jaén* and *aragonés* varieties bear much fruit when trussed. Grapevines to be trussed should be planted when they are seedlings, because then they are more certain to grow.

When trussed grapevines are planted in areas where grain is also grown, special considerations need to be taken into account. In Italy, for example, the following counsel applies. For grapevines planted in grain-growing areas, there should be a forty- to fifty-foot distance between trees, allowing the sun to shine on the grain field from dawn to dusk. If grain is grown on only one side of the field, each row should be separated by a wide row aligned in such a way that the sun shines on the grain and a fifteen- to twenty-foot distance between trees invites the vines to grasp them as if they were dancing together. By contrast, for grapevines planted in areas where grain is not grown, a twenty-foot distance between trees is sufficient.

To truss a grapevine in a tree, first tie two or three strings around the tree trunk at the approximate height of a man, so the vine shoots can grasp them and grow, supported, up the trunk. As they do, no part of the grapevine should be hung, as this causes damage at the tying site. Instead, several strings should be placed on the tree branches, allowing the vine shoots to grasp them. Then any long, high, uneven branches should be removed to prevent the grapevine from towering overhead, where its branches could obstruct pruning and harvesting.

Not all trees are suitable for trussing grapevines, however. Trees with extensive shadows and thick leaves should be avoided, because grapevines do not grasp such trees and the grapes, due to the shadows, fail to grow well, lack flavor, spoil quickly, and do not produce good wine. I also recommend not trussing grapevines in fruit trees, because they often spoil. If grapevines are already trussed in fruit trees, they should be pruned before the trees bud, preferably during the waxing moon at the beginning of winter, because this is when tentacles grasping the trees begin to spoil at an increased rate. For easy removal, vine shoots should be pruned after the tentacles have spoiled.

It is preferable to truss grapevines in trees with flexible wood that will not crack easily while the vineyard is being maintained. White poplars or willows are best for trussing in moist lands because these trees flourish under damp conditions, while black poplars or ash trees do better in dry regions. For vineyards planted on hills, ash trees are the best of all. Moreover, Columella advises that the tree in which a grapevine is trussed should be only slightly older than the grapevine because if it is much older it can smother the grapevine or stunt its growth. I believe, however, that the tree need not necessarily be older because vines usually take root and flourish regardless. Often, though a grapevine planted at the base of a large tree may perish, another planted in its place will thrive.

If the tree and grapevine are transplanted at the same time, it should be done in accordance with the following method. Both holes should be dug many days in advance, then the tree should be planted as well as possible and its roots covered with a little dirt. If the region is hot, with intense sunlight, plant the grapevine facing the north wind so the tree will protect it from too much sun. If the region is cold, with a constant north wind, plant the grapevine facing the noonday sun so a portion of the tree will provide shelter. In a temperate region, the grapevine should face west, or whatever direction best suits the plant. In any case, the hole for the grapevine should be dug about one and a half feet away from the base of the tree so neither will be in danger of perishing or having its growth stunted; in addition, the hole should be as deep as the land allows, providing a solid base to ensure the plant's strength and fruitfulness. If the land is moist, two or three baskets of stale manure should be combined with twice as much soil at the bottom of the hole, along with rocks, and the soil should be well compacted. The topsoil, by contrast, should remain loose, creating a concave hole that allows for the absorption of water.

As the tree and grapevine grow, ensure that the grapevine does not climb the length of the tree trunk in the first year, because then it is likely to become unmanageable and thin, straying in all directions and losing vitality. Ideally, it should grow to a height of eleven feet or slightly more in its first year, its growth both stimulated and controlled by the fact that it has been tied into three sections and pruned three times. If instead it is to climb quickly, its lower tips should be removed, leaving only two or three tips on its uppermost part to foster rapid growth. A vine that has been tumbled, or inverted and planted underground, can endure any height its first year. Finally, the soil around every new grapevine must be dug up or plowed in the morning or early afternoon once a month, beginning in March.

HOW TO PROP AND TIE GRAPEVINES

Once grapevines are planted, they need "tutors," just as children do, to protect them and guide them. Chestnut trees in close proximity make ideal tutors, since they survive for a long period without rotting. Ash, pine, juniper, and almond trees are also suitable. Neither hazelnut, laurel, nor pistachio trees should be used, however, because they have a strong odor and attract thousands of worms and lice.

A prop, or tree limb, should be dry and unbent, so the grapevine it guides is trussed straight. The tree's branches should be placed close to the ground, allowing the grapevine to grasp them with its tentacles, thus securing the vine shoot and preventing movement from severely harming it. To stop it from catching the wind, it should not be very long—certainly no more than four palm lengths above the ground—and should be at least as thick as a lance's handle. In cold regions, farmers recommend placing the prop toward the north wind, thereby protecting the grapevine from the cold. If the region is hot, it should be placed toward the noonday sun, protecting the grapevine from the heat. If the land is temperate, the prop can be placed in any of the four directions. Of course, in the event the new grapevine was planted near a tree, there is no need for a prop unless the tree is too thick for the vine's tentacles to grasp it securely.

Regarding the type of ties to be used, hard materials should never be utilized as ties, because they can stunt the vine shoot's growth. If a hard material must be used, something soft, like lettuce or an old linen cloth, should be wrapped around it to ensure that the tie does not cut the vine. In addition, the tie should never be placed on the tip of the plant. Moreover, if the vine needs to be tied for a long period of time, the tie should be moved every year and placed on the older section of the vine, where it will do less harm. If the grapevine or vine shoot is tall, it should be tied to the tree or prop in two or three different places to reduce the potential for damage. It should never be suspended, but rather supported by the tree. Finally, there are two times when tying can be done without causing damage to the grapevine: before the tips begin to sprout and when the unripened grape bunches are firm.

INITIAL INSTRUCTIONS FOR THE PRUNER

Since the first two or three prunings determine to a great extent whether or not a stock will succeed, the best pruner available should be hired to prune a new grapevine. As was discussed earlier, certain tips should be left on each vine shoot so new shoots can grow, thus forming the grapevine. Although many people prune everything that sprouts the first year, it is preferable to allow the branches to grow until time for the second pruning, only cleaning them initially and thereby allowing the grapevine to gain strength. Once more branches have grown, the pruner can decide what to remove and what to leave. Another option is to plow the plants slightly and hoe all that has sprouted, stimulating the growth of new branches and vine shoots.

Before preparing to prune, it should be noted that in humid, rainy lands and also on plains, grapevines grow taller than they do on hills, in sandy areas, or in loose soil. Prune with this in mind, ensuring that the resulting grapevine height is suitable for the type of land. In some areas, grapevines can be reduced by two or three palm lengths, and in others by one-third or one-half because being on dry land or in places where the water is quickly absorbed, they will not spoil.

Grapevine varieties that grow sprawling over the ground should ideally be planted in sandy soils. Lacking form, they must be trussed on stakes of a suitable height to prevent spoilage, and their branches placed in a cross or star shape so the base carries equal weight on all sides. These varieties should have at least three boughs for support, since a grapevine with only one or two boughs carries all the weight on one side, causing the base to break. Also, the greater number of boughs the more shoots will be generated and the more fruit they will produce.

In hot or dry regions, grapevine branches should be spread sufficiently wide to protect both the base and the grapes from overexposure to the sun. In cold or humid lands, the branches should be gathered more closely, permitting the sun and air to dry any moisture and the oxen or mules to plow without breaking the vines.

TIPS ON EXCAVATING

Excavation of the vineyard is a prerequisite for effective pruning. In terms of vineyard care, to excavate—from the Spanish word *excavar*—is not to unearth or dig up the roots or bases of grapevines, but rather to carefully loosen and weed the soil surrounding each base, leaving a hole around it. The purpose of excavating in a dry region is to provide an area where water can gather during the summer, and in a humid region to give the grapevine more air and sunshine. This task must be undertaken every year, particularly for new grapevines, which tend to grow shoots on the ground's surface; if these are not cut, they can eventually sap the strength of the roots underneath, possibly killing the entire vine. An annual cutting is usually sufficient, since the surface shoots tend to be burned from the heat in summer and the cold in winter. Avoiding excavation altogether, through shallow planting, is not an option, for grapevines lacking deep roots do not gain sufficient nourishment or moisture, and consequently produce fruit that is undersized and fails to ripen properly.

When excavating, a sharp tool should be used to cut the shoots as close as possible to the grapevine, at a palm's depth beneath the surface. They should never be pulled out, because this both damages grapevines and either fails to remove shoots or removes too many—a process similar to the removal of hair, which can be either pulled out or properly cut with scissors. The shoots should be cut on clear, calm, hot days, and they should be protected from the north wind, which is detrimental to all fieldwork. It is best not to cut them in the spring, although in fifteen to twenty days they will dry somewhat. By contrast, in October or November they can be easily removed and cut since they do not bleed at that time.

There are two ideal times for excavating. In cold regions, excavation should be undertaken during or after February. In hot and dry regions, it should be done after the harvest. Even if plants remain in this state all winter, it causes no harm; in fact, they benefit from being saturated with water. In dry and hard lands, excavation is done after the first rain, when the soil is soft. In any case, this procedure should be done before pruning, which will remove all the shoots, seedlings, and rubbish that grow underneath—a task that can only be accomplished if the grapevine has first been excavated, or at least hoed. And it is best done in loose soil, which can be worked whether it has rained or not.

The excavation should be deep, wide, and carefully executed, so as not to damage shallow roots. To prevent the evaporation of moisture, the hole should be covered with soil when the weather begins to turn warm—preferably, in dry or hot regions, before the vineyard sprouts. In hot, dry, or cold lands, the entire bases of the grapevines are often covered, while in humid regions the soil remains lower to prevent spoilage of the grapes. I maintain that covering the entire base with soil is harmful for a grapevine that begins to develop early, because raising the soil around its base makes the roots grow high. It is best if the soil remains level rather than piled on the base, since then it can stay moist without causing the grapevine to spoil or resulting in other possible damage.

An excavation method suitable for both cold and dry regions that allows grapevines to absorb water during winter without being harmed by the cold is the following. After covering the grapevines with soil, a ditch is dug around the pile, forming a crown. This is a good technique to use during summers when the soil is very dry, provided the soil cover remains low at the base of the grapevine.

FURTHER INSTRUCTIONS FOR THE PRUNER

There are two propitious times for pruning, although these may not be applicable to all lands or all vineyards. The first appropriate time is just after harvesting, while the other is at the beginning of spring, around February or March. The most suitable time for a particular vineyard can be determined as follows. Grapevines pruned before winter neither bleed, which means that their substance is not depleted through the cuts, nor lose moisture from the cuts.

Hence, pruning before winter is best for older, thinner grapevines; for those planted in loose, airy, sandy soils; and for those on hills where they receive little moisture. In cold regions, pruning before winter is not necessarily infallible, because the cuts tend to burn. Nonetheless, when pruned before winter the vine shoots seldom split, for they are still tender. Moreover, agriculturalists provide this notable information: when grapevines are pruned before winter, they produce more wood, and when pruned immediately after winter or later, they bear more fruit.

To assess whether particular grapevines can be pruned after winter, the owner of the vineyard or the pruner must first evaluate their strength. If a grapevine is thin, old, or afflicted in some way, and needs to form anew, the pruning should be done early, allowing the necessary time for it to produce more wood. Vineyards located in very hot regions should be pruned before winter; those in very cold areas should be pruned during March; while vineyards in temperate lands with mild winters can be pruned either before or after winter. Vineyards exposed to the sun but sheltered from the north wind can be pruned before winter, even if located in cold regions. Vineyards in very hot regions that are exposed to the north wind should be pruned according to the specifications provided for cold regions. If these vineyards are pruned before winter, they may freeze with the first north wind.

Grapevines should never be touched in freezing weather, neither with a steel instrument nor anything else, because they are very tender and break like glass. Hence, avoid walking among grapevines during the entire month of December, due to the severe damage it could cause. And when pruning in January or February, make sure the day is clear, warm, and without wind, so the stocks are well thawed and not brittle.

Come spring, it must be remembered that to grapevines, pruning is a form of "punishment." To lessen the damages—in this case, to reduce the bleeding—spring pruning should take place during the waning moon, unless the grapevines are in very humid regions, or are robust and concentrate their strength in branches without fruit. In cold regions, pruning is best performed during the waning moon in March or in February if temperatures have already warmed. Pruning during the waning moon is not applicable before winter, however, because at this time grapevines do not bleed.

Further, the pruning process differs somewhat according to when grapevines are pruned. Pruning before winter, which should be performed when grapevines have just lost their leaves, must be done as quickly as possible so the cuts have sufficient time to harden before freezes cause harm to vine sections. In addition, it is best to cut all twigs from the stocks, which removes some of the weight and, most importantly, prevents tearing. If pruned in spring, the cut twigs bleed far more than a cut vine shoot. For grapevines pruned during spring, the process should be started as soon as the weather begins to warm, and be completed quickly, before the vines begin to bud. Pliny maintains that in ancient times pruners were deemed negligent if the vineyards were not pruned before the cuckoos arrived.

Regarding pruning requirements, skills, and tools, the pruner should know both the vineyard's soil and the varieties of grapevines in it. Because of this acquired knowledge, the same person should be in charge of pruning year after year; changing pruners every year is detrimental, even if they all prune well. Further, the pruner should be very strong and capable of cutting a vine shoot with a single stroke, so as not to break the vine. The pruner should also carry a good dagger to cut dry and irregular boughs, as pruning knives are not adequate for such tasks.

Before pruning a grapevine, the pruner must carefully examine the vine in its entirety, because one side may require a certain type of pruning while the other may need to be pruned differently. In addition the pruner, upon determining which branches should be allowed to grow and which should be removed, must have two main objectives in mind: to produce better fruit and to maintain optimal health of the branches. A variety of pruning knives should then be used, according to their different purposes and the types of grapevines on the land. For efficient work, each pruning knife should be sharp and, according to Columella, have a hooked tip for cutting roots, fibers, and above all, vine shoots that can otherwise be reached only by bending them, which greatly damages the vine.

The pruner must ensure that from the time the shoots are small, the shape of the stock corresponds to the nature of the land. In humid regions, the stocks should be high and slender, while in dry areas they should be low and extended. The same holds true for hot regions, so branches can shelter fruit from the sun, avoiding desiccation. Similarly, vines in areas that must be plowed should first be gathered to prevent damage. Then too, pruners often thin grapevines at the base to promote rapid growth, but vines should be allowed to grow incrementally or they will generate branches that are much thicker than the base, rather like a large building over a small foundation. Vines pruned in this manner have diminished strength and longevity, bear little fruit, break easily, and are quickly uprooted by the wind. Thus, to ensure a proper shape from the beginning, it is important to employ a good pruner during the first four or five years.

Grapevines are more amenable to pruning than any other plant and accepting of whatever shape they are given, but to ensure that they carry weight equally over the entire base, the pruner should make certain that each vine has five boughs, or four boughs in the shape of a cross, but never less than three. This produces vines that are more attractive, firmer, and more fruitful. Grapevines that carry weight unevenly on one side are commonly worm ridden, emaciated, and prone to damage from cold or sun. They must be hoed from below, allowing them to generate new shoots from which fresh stocks can grow. Sometimes a new vine shoot sprouts underneath, called a *tornillo* (screw), from which a vine can grow anew after the growth above has been cut away. The method can also be applied to old, emaciated, or unmanageable grapevines, allowing them to restore themselves from beneath. In addition, the heads can be tumbled.

For optimal pruning, it is helpful to plant each grapevine variety separately in the vineyard, allowing for the easy identification of each one so it can be pruned at the appropriate time and in the correct manner. Some must be pruned early, while others should be pruned when they bud. For example, delicate grapes such as *alvillas, castellanas.* and similar varieties that bud early should be pruned before those that bud late (such as firm grapes like *jaénes* and *palominas*). If different varieties are intermingled, there is no choice but to prune them all simultaneously, which is deleterious because it might be late for some and early for others. In addition, because a fruitless year leaves stocks thin and in need of being constrained to prevent disease or desiccation the following year, it is beneficial for the pruner to know which types did or did not bear well the previous year. In general, the newer sections should be pruned as much as possible so they can produce better branches and fruit, while all sections that are dry, old, decayed, ant infested, worm ridden, sterile, or overgrown must be removed. If vine shoots in the older, damaged section are to be used to grow a new grapevine, sometimes called *mugrón* (sprig), they should be transplanted in another area. Simultaneously, remove wayward vine shoots and buds and vine shoots born in clusters.

This pruning method should be used for young and old grapevines alike, but for different reasons. The young ones should be cut back so they may flourish, and the older ones so they do not perish. Very old vineyards should be tightly pruned; the younger ones can better withstand weight since they have more strength. If the soil is rich, or exceedingly moist, the pruning should be more extensive than in thin, loose, and not very fertile lands. Vines grown on plains should be pruned more thoroughly than those on hills, and pruning should be more vigorous in valleys than on plains. If a stock lacks proper form, it should be hoed from below, and two slits made in the greenest section, which will produce a vine shoot capable of reshaping the stock. If a vine shoot grows from below, its tips should be removed and transplanted elsewhere, including close to the vine, if desired. Alternately, the vine shoot can be covered with soil and, with some of the tips removed, grow quickly.

When pruning, it is important to squeeze the vine shoot very tightly, without splitting it. In fact, most shoots will split when cut if this method is *not* used, causing permanent damage, since cold and heat enter through the torn area, stopping them from mending properly. It is said that for vineyards located in cold regions the cut should be made facing the noonday sun, thus preventing freezing, while for those in very hot areas the cut should be made facing the north wind, preventing excessive sun from penetrating and harming it. If the land is temperate, the cut can be made in any direction, although most pruners prefer to have it face the sun. I do not put much faith in this method of cutting in a particular direction, however, because I do not believe the presumed impacts are possible.

Each cut's positioning is also significant. Because the vine emits a substance that burns the buds it touches, causing damage at the vine juncture and thus preventing strong branches from growing in that section, farmers pruning in spring should cut at the nodes, or at least never closer than half the length of the section. Cutting at the nodes is easier because they are similar to joints, and preferable because they are far from the buds. In addition, the cut should face downward, as an upward-facing cut will most likely cause the substance to drip on the vine shoot below, burning its buds.

Finally, grapevines thick with buds require extensive pruning to retain their vigor. Thin grapevines, regardless of their age, do best when twigs are removed from them. If left on the vine, they should appear only in irregular sections, so the following year they may be removed along with the elongated branches they grow on. Whenever possible, such grapevines should carry their twigs on top, because there they are less likely to spoil. For grapevines that are strong and growing in good soil, twigs may be left on the vine as long as the tips are trimmed, which is preferable to leaving them intact. Although vines produce fewer bunches when their tips are trimmed, their grapes are far superior. Ten well-pruned and well-trimmed grapevines generate more fruit than twenty unkempt grapevines. Although some farmers today may value little or none of this information because it is not in current use, to them I say that in seeking a model we must turn not to what is currently used but rather to what should be used.

SECRETS ABOUT GRAFTING

Many who prune in the spring turn next to grafting—inserting scions in trunks or branches of stock that then infuse them with sap—especially in the latter part of March. Grafting earlier, in winter, is not generally effective, due to the freezes and rains that significantly damage the scions and stocks. If one wishes to graft in winter, it should be performed a month after the harvest in warm areas and protected places, particularly those sheltered from the north wind. Although winter grafts are sometimes successful, I do not recommend grafting at that time, since grapevines do not bleed then and, more importantly, the vine shoots are not yet properly seasoned. In colder regions, grafting can be performed as late as April, because hot and humid weather is favorable for the procedure. In fact, grafts performed at this time generally result in better growth than grafts made at any other time.

Whatever month is ultimately decided on, all grafts should be done during the first phase of the waxing moon, because that is when they are most likely to take. If grapevines are robust, grafting is best done when the weather is calm and clear of wind and rain, preferably during the waning of the day—that is, in the afternoon or evening. If grapevines are not robust, they should probably not be grafted for another year or so. Of course, even then not every grapevine is suitable for grafting. Scions that are decayed, or too old, for instance, will not take. A good foundation ensures that vines will take better and be more fruitful.

The tools required for grafting are as follows. First, one needs a very thin, small saw that makes fine cuts and so is less likely to split the vine; a thin knife or butteris to smooth the saw cut, as a farrier does with a horse's hoof; a knife to split the trunk; and a long, smooth wedge shaped like a thumb and of comparable thickness. Some farmers make the latter from bone, because it is smooth and leaves no residue, while others use dense, hard, smooth wood like oak or box tree, and still others use a steel chisel, which is not advisable because it leaves rust that often prevents the graft from taking. Additional grafting supplies include sticky white or red clay, easily molded cow manure, and strips of cloth or old rags.

There are six basic grafting methods. The four fundamental ones, which are generally the most beneficial and successful, are *de mesa, de barreno, pasar,* and *empalmar,* while the other two are *de yema* and *de juntar.*

The grafting method known as *de mesa* is performed as follows. If the grafting is done in a cold region or a grapevine is particularly robust, it should be cut a palm's length above the area where the graft will be made. This expels some of the vine's sap, preventing it from engulfing the grafting scion. If one prefers, two or three gashes can be made above the area to be grafted instead of actually cutting the plant and allowing it to bleed in that area. These procedures apply whether the graft is made in the trunk or base of the vine, although neither one is necessary if the graft is located on a higher bough, because boughs do not contain as much sap. In addition, some of the vine shoots can be removed, but I do not consider this a good practice since if the vine is cut, it should not be done before grafting. The vine should, however, be hoed around the area where the graft will be located, and the trunk tied tightly under the incision to ensure it will not split. The incision itself will be made as low on the vine as possible, preferably underground, provided it is not too deep, because this part of the grapevine is most tender and the place where grafts take best.

The vine shoots for scions should be taken from the best part of the vine—one that is smooth and healthy, with thick, dense buds. The selected shoots should be thin, for these are more solid than thicker shoots; and because vine shoots are hollow, those selected for grafting should be as solid as possible, particularly when using this method. Some of the best vine shoots for grafting are those generated from others, the "grandchildren," because they are firmer. A two-year-old "grandchild" is ideal since it has almost no pith, although a graft with a year's growth can also work. In any event, it should contain only three or four buds, all of which are healthy. Once selected, the vine shoots should be cut past the midsection of their length and the tips discarded. The cut scions can be preserved for a few days, submerged in a pot of water.

To make the grafting incision, dig underground to a section of the trunk that is healthy, green, smooth, and not prematurely old, decayed, or worm ridden. Then cut carefully three or fingers' length down the center, without splitting it. The incision should be straight from top to bottom and smooth, while remaining somewhat elevated in the middle so sap runs to the grafted scions.

After the trunk is cut, the wedge should be inserted high in the center of the incision, as if it were growing from the vine. A selected scion, with its three or four buds, should then be trimmed sufficiently but no more than is required to enter the trunk, leaving the pith untouched. (Another trimming method calls for all branches and most of the pith in the chosen grafting area also to be removed.) To begin the insertion, one bud should be placed inside the graft or just outside of it, with the other two or three remaining outside. The scion should then be trimmed again, enabling all buds to be gently inserted. At this point, the skin of the scion and the trunk will be fitting together so well they are almost indistinguishable and the remaining portion of the scion will be parallel with the flat section (*mesa*) of the grapevine. If the trunk is thick enough, as many as three more scions may be inserted along the slit, after which the wedge must be removed, allowing the trunk to tighten around the graft. (Some agriculturalists, in lieu of making a large incision, cut only where a scion enters the trunk, one after another, filling previously empty areas.)

To tighten well, the trunk should be tied securely if the grapevine is thin or, if thick, left untied. So that neither sap nor rain penetrates the delicate areas, a covering should be placed over the graft as follows. First, well-crushed dry leaves should be pressed gently around each scion and just covering the incision, to stimulate the production of gum while protecting the graft. Clay should then be molded around the cuts, and the scions tied with rags to prevent them from moving. Also, a trench should be dug in the soil surrounding the scions so rainwater will not collect there and burn them, and two or three props should be placed firmly around the graft to protect it from movement caused by the wind.

Finally, if the soil is moist or the grapevine strong and previously uncut, a small slash should be made below the graft, allowing it to bleed moderately without engulfing the scion—a task that is unnecessary if the grapevine is in loose, dry, sandy soil or is thin. If the soil is dry, water the vine a few times without letting the water touch the graft. If the graft is not located underground or very low on the vine, gather soil around it or, alternately, cover it with clay, cow manure, and rags to protect it from winds, cold weather, and sun. This method can also be used to graft other plants with small seed into the same vine or others. Regardless of the types of plants used as scions, submerging the graft the following year will strengthen the juncture and make the new growth more fruitful.

The second grafting method, *de barreno,* is very simple, requires few tools, and proves to be highly successful. First, a downward-slanting hole is made in the trunk of the grapevine, preferably at a branch juncture, reaching all the way to the pith of the midsection. Columella maintains that an effective instrument for making this hole is what he calls a *terebra gallica* and others call a *taladro*, a type of auger. But this tool can leave sawdust-like remnants inside the vine, causing the graft to perish. In my opinion, a better tool is one made of iron with a hollow half-circle resembling a fingernail, which some call a gouge (*membriqui*). This tool cuts quite effectively when well sharpened and turned like an auger, removing everything it cuts and leaving no residue.

After the hole is made, it should be cleaned and allowed to cool so the water will not burn the selected scion when it is inserted. While the hole is cooling, measure its depth with a small twig and trim the scion accordingly, scraping evenly so it will fit well. If the trimming removes only a thin layer below the skin, the scion, once inserted, will swell and prevent water penetration, even during a rainfall. Then insert the scion and cover the graft with clay, a rag, and a prop, in keeping with the foregoing indications. The grapevine can also be hoed by using an auger to make a hole in the center of the trunk, a good technique to use when grafting a vine with black poplars, mulberry trees, apple trees, pear trees, or other varieties of trees.

The grafting method known as *pasar,* although somewhat difficult and time-consuming, is the most consistently successful of them all, providing for the grafting together of even difficult trees, such as fig and olive. *Pasar*'s extraordinary success rate is due to the fact that it permits each scion to be nourished by its mother plant until it takes well. A grapevine should be planted next to the grapevine or tree that will be grafted. This new plant should be well tended until it begins to thrive, at which point a scion about two years old but no older should be selected and a hole made through the grapevine or tree branch to be grafted. The grafting branch should be fresh and smooth, with enough strength to fuse quickly. It should not be very thin, and the hole or incision should penetrate the center of the pith, allowing the scion to fit comfortably, without injuring its buds. To facilitate its access, I recommend trimming the branch through which the scion will pass angled upward, as acutely as possible, as it moves through the hole so that more of it will stay inside the branch and therefore receive greater nourishment. The grafting branch should then be well tied and covered with clay and other protective coatings, as explained previously. When sufficient time has passed for fusion to take place, the graft should be cut from underneath, very close to the trunk of the tree so it can be better incorporated into it, then well covered. This method allows fruit to be altered into different varieties at different times. For example, grapevines grafted with cherry trees will, through this technique, produce grapes at the same time the cherry tree renders its fruit, which ordinarily occurs much earlier. Mulberry trees too can be successfully grafted using this method.

A very similar grafting method with a high success rate that can be used when two grapevines are in close proximity is the following. A hole can be made in the bottom part of one vine's trunk and a vine shoot from the other scraped and inserted. Alternately, a hole can be made in the main root of one vine, through which the properly scraped vine shoot can be inserted, ensuring that it fits well. In both instances, the shoot will take root in the direction opposite from that in which it was growing while grafted. Once the shoot is successfully thriving, it can be cut from the mother vine.

The fourth basic grafting method, *empalmar,* is used exclusively for grapevines and is an excellent means for moving a grapevine to a new location or altering it to a better variety of grapevine. This method is done in the following manner, working as before with a vine shoot that is more than a year old, and ideally two years old, so it will be fairly solid and a scion from a good grapevine variety that also is two years old and thus somewhat solid with relatively little pith. A hole approximately the same size as for tumbling a vine shoot should be dug at the base of the scion's stock. Then an incision is made down the center of the vine shoot, where it is most green. The vine shoot, still attached to the mother plant, should be split about the length of two or three fingers and cut a palm's length from the point and a bit more from the head. The area that will enter the split vine shoot should be appropriately trimmed, without touching the pith. The juncture should be at an upward-slanting angle, allowing the two segments to join tightly. Once well joined, they should be tied, and a split cane placed over them to hold them together. The graft should then be covered with clay and rags and placed very carefully in the ground, ensuring that the juncture is not disturbed and that three or four buds remain on the exposed section of the scion. After one or two years, the grafted vine shoot can be cut from the mother plant, transported to its new location, and gradually covered with soil. This is an excellent grafting technique because it combines the qualities of a vine shoot, which grows quickly, with those of a graft, which creates a better vine.

A variation of this grafting method is performed as follows. A fairly thin grafting vine shoot should be cut, leaving approximately four fingers' length of the older section from the previous year. A hole should be made in a two-year-old mother plant, extending a little beyond the pith so the vine shoot can be inserted. The point of the vine shoot should be trimmed to the same width as the hole, and scraped until it can be properly joined. It should then be covered with clay and placed underground as described earlier, leaving only the tip exposed. If the objective is to have many grafts in a single stock, tumble the grapevine in a hole, removing each point as if it were a vine shoot and thereby allowing for an *empalmar* graft on each one. This graft is best performed a year after the grapevine has been pruned, so the vine shoots are more solid. Ideally, the scions should be cut during the waning moon and protected from harm and wind, while grapevines should be grafted during the first phases of the waxing moon.

The grafting method known as *de yema*, which can only be performed in the spring, when the buds are plump and vine shoots perspire copiously, involves the following. When the tips are plump with healthy buds and before leaves begin to grow, remove an entire tip with a very sharp knife point. Then, from the best section of the grafting vine shoot, remove a bud using the same technique and replace it with a tip from the scion, ensuring a tight fit. Some farmers precede the replacement with a drop of honey to enhance the grapevine's adhesive capacity; however, it should be an almost imperceptible amount, because honey burns. A better technique is to take a drop of gum called tragacanth (*alquitira*), thin it with water, and place it on the juncture. My preference is to bypass adhesives altogether, because grapevines secrete a significant amount of sticky sap that serves as a more effective adhesive than any other substance.

Alternately, a variation of this method is performed as follows. When the bud has bloomed and grown leaves, remove it with a knife point, letting the leaves remain on the vine shoot. In its place insert an intact bud that has not yet bloomed. Then mold a little cow or goat manure around the juncture.

The grafting method called *de juntar,* performed more for aesthetic reasons, is better suited for flower and fruit gardens than for vineyards situated in fields. Its principal objective is to combine many different grape types and colors in one bunch. There are two variations of this method. In one variation, vine shoots of different grapevine colors or types are joined, balanced and well paired, close to where they originate. Incisions are made among the buds, ensuring the vibrant health of those that remain. They are then joined at the incisions so they appear to be one, well tied, and covered with clay. The buds must stay outside the tie. If the buds are long enough to be buried, it is best that the tips stay exposed. After they are fused well enough to appear as one, they are cut from the mother plant. The section of the trunk ordinarily placed underground is covered with soil, allowing it to grow roots.

Afterward, the tips are unearthed and cut at the junctures. The vine shoots born at the junctures will produce diverse bunches.

While the first variation of this method merges two lineages of grapes, the second allows for the fusion of four, five, or even more varieties and is performed as follows. Take a clay cylinder having the length of two palms and the thickness of a hoe handle, and split it down the middle from top to bottom. On the split cylinder, place four or five thin vine shoots that, preferably, have taken root elsewhere after being planted together. Two-thirds of their length should be inserted into the cylinder, and the two should be well tied, preventing separation. This allows the vine shoots to merge as they grow. Some farmers insert the vine shoots into a bovine shinbone, but this cannot be done without damaging the buds and the vine shoots. Instead, they should be planted in the soil together with the cylinder, leaving only the tips exposed. If using seedlings, separate them individually so they take better. When they are well fused together, the cylinder should be loosened. The vine shoots, which surface as the stock is being hoed, should be cut where they are well merged, and again covered almost completely with soil. Columella maintains that tendrils will sprout from that juncture and the better ones should be allowed to grow, as they will carry bunches of different colors and varieties. Some agriculturalists recommend scoring the vine shoots twice or thrice in this area, which will promote budding, although scoring the branch or the root will not lead to budding.

Finally, Albumaran Abencenif describes another method that does not require cutting the vine shoots from the grapevine. When pruning a white grapevine near a dark grapevine (or any other color or variety), the best, slimmest, most flexible vine shoot on each should be removed and grafted together, cutting the ends evenly. Buds should be merged with other buds, and vine shoots with other vine shoots. These should be covered with clay and sprinkled with river water every three days. After two years, they should be cut from the mother plants and transplanted elsewhere. It is said that these will bear white and dark grapes in the same bunch.

This method also works for grafting three different varieties and colors of grapevines. If a farmer takes one vine shoot from each and cuts them carefully, taking care not to harm the marrow, they can be merged. With buds sufficiently similar, the plants fuse into what appears to be a single vine shoot. These should be tied with sedge, cords, or willow reeds, leaving the buds outside the tie. They should then be well covered with cow manure and good clay, and placed in a hole two palm lengths deep, or a little deeper, with no more than two or three of the tied buds exposed. They should be watered as much as necessary until they successfully fuse. It is said these buds will produce vine shoots carrying all the different grafted varieties in one bunch. If they are to be transplanted, they should be at least two years old to ensure success.

Farmers interested in grafting different types of grapevines to produce amalgamated bunches should graft them all in the same stock using the de *mesa* method, especially if the vine shoots are on the grapevine, which is easier to do than if they are on separate grapevines. Abencenif asserts that if grapevines are grafted with myrtle (*arrayhán*), a leaf will grow between every two grapes. These can be grafted using either the *de barreno* or *pasado* method, or even the *de mesa* method; cutting the myrtle branches, so they take better; then transplanting them after two years.

METHODS FOR PRODUCING MEDICINES, FRAGRANCES, AND SEEDLESS GRAPES

Medicines and fragrances can be successfully grafted to grapevines so the plants take on their attributes, and the resulting grapes, raisins, and wine can be used for a variety of purposes. Palladius explains how to graft treacle into grapevines to produce grapes, raisins, vinegar, and grapevine ashes with the same attributes as the former for healing poisonous bites, as well as for other maladies and wounds, as follows. When planting a vine shoot, score it three or four finger lengths in the center of the pith. Carefully remove all the marrow and replace it with the best treacle available. Then tie it well with sedge. Some people insert the vine shoot head in a squill, because this preserves its green color for an extended time, until it takes. Abencenif recommends irrigating the grafted shoot with a little treacle diluted in water every eight days until it has taken. Using the same technique, the vine shoot filled with scammony grains, as well as the wine, grapes, and raisins, will effectively cleanse the abdomen. If filled with celery seeds, it aids in sleeping. Other medicines can also be made this way. Nonetheless, I would not graft fragrances and medicines using the technique Palladius recommends. I believe that before a grapevine grafted with this method takes, all or much of it will perish due to the fact that it is underground, surrounded by moisture.

The following method for grafting any fragrance, flavor, or medicine is easier, quicker, and better. When a grapevine is relatively young, meaning five or six years old, and pruned before winter, the graft should be performed. At a juncture below the shoots left on the vine, an oblique hole should be made, reaching the heart of the grapevine. It should be well cleaned, and musk or amber should be inserted to provide a pleasant aroma, or saffron to enliven the heart (although wine makes people sufficiently happy in and of itself), or whatever one prefers. A generous amount should be inserted to fill the grapevine adequately. Then the area should be covered with a very tight-fitting and strong wedge to prevent the loss of moisture, and coated with clay. This should be performed during the first phases of the waxing moon in the spring, using only a young grapevine before it begins to bloom. With this method, any fragrance, flavor, or medicine can be grafted.

Further, Palladius describes another technique for grafting fragrances and medicines as follows. In a new pot, dissolve a little soil in approximately one liter of rose water or any other fragrance. The soil should be fresh, never worked or fertilized—what some refer to as virgin soil. Denser soil is better than loose soil for this purpose. Place the vine shoot heads in the pot and bury it. Cover the pot completely, so there is no exposure to sun and nothing can fall inside. When the buds swell slightly, they should be planted in the preselected location. Any fragrance or fortifying medicine can be dissolved in the aforementioned water. To fortify the plants until they begin thriving, the roots should be watered from a cylinder containing the same water. In my opinion, it is better, easier, and more reliable if the grafting scions are placed in the water, rather than the vine shoots for planting.

To produce seedless grapes, which are excellent for making raisins, when pruning in the spring select a healthy vine shoot that is new, thin, and flexible. If it is from a vine suitable for transplanting, it will be even more reliable. Leaving the vine shoot on the grapevine, split it evenly through the pith, along the area where it will be inserted into the ground. Carefully remove all the marrow or pith, without going beyond the aforementioned point on the vine shoot, then replace all the pith. Tie the split vine shoot, except for the buds, with sedge and cover it with calf manure. Then bury the vine shoot just beyond the point where the cut was made. Some farmers insert the head of the vine shoot in a *cebolla albarrana,* or squill, as mentioned previously, to both maintain its freshness for an extended time and foster rooting.

This form of grafting should be done in humid regions with sufficient moisture, or where there is water in close proximity, because after removing the marrow, vine shoots need to be strengthened by water, even if it is only the artificial moisture of irrigation. Some farmers bury a well-covered horn or a new unglazed earthen jug full of water next to the transplanted vine shoot to foster better water seepage and help season the plant. Moreover, sufficient shade should be provided.

In my opinion, all these factors combine to produce fruit more quickly and keep the vine shoot from perishing. In time, vine shoots will sprout from the buds where the marrow was removed, carrying seedless grapes. Hence, the grapevine will have some grapes with seeds and some without. Anyone who wishes to undertake these grafting experiments may achieve interesting results, but they should make numerous attempts since not all plants thrive.

ABOUT GRAPEVINE DISEASES AND THEIR CURES

Spring is the most suitable time to remedy problems associated with grapevines. There are two types of diseases that afflict grapevines: some are internal, akin to human diseases, while others are external, such as knife wounds and contusions. One internal malady causes vines to produce grapes that do not ripen properly. Curing this disease requires the following process. A deep hole should be dug, and a mixture of a substantial quantity of vine shoot ashes and putrid human urine combined with an equal portion of water should be poured on the roots. A small quantity of vinegar diluted in water, together with grapevine or vine shoot ashes, can also be used on the roots. In addition, covering the entire grapevine with a vinegar paste is beneficial. These treatments should be applied several days before grapevines bud and the weather turns hot.

If a grapevine has desiccated sections, is ant infested, or worm ridden, all the dried and diseased areas should be scraped away so the vine can grow again. Then unsalted olive juice, which repels ants and other small insects, should be kneaded with clay and molded over the entire scraped area. If olive juice is not available, several wild olive tree leaves (or ordinary olive leaves if wild olive trees are not accessible) can be boiled and clay can be kneaded into the water-leaf mixture, then molded over the vine before covering it with soil.

If a vine bleeds excessively when pruned, unearth its roots and score the fattest one with a knife. Abencenif recommends that when a grapevine bleeds excessively, the roots should be completely unearthed and the largest vein cut. A few days later, it should be covered with clay and then soil. In my opinion, there is no better remedy than tumbling old, sick, or decayed grapevines. The old becomes new, the sick healthy, and the sterile fruitful. In addition, a vine can be hoed to generate new branch growth or create a new grapevine; it can also be grafted.

Grapevine leaves turning yellow or red before their time is a sign of frailty. To help strengthen such plants, make a hole in the root with a chisel, auger, or drill; insert a wedge of any type of wood to prevent its closure; and cover it with soil. The wedge should fit loosely, and the area should be irrigated every few days with salted water. If leaves or fruit fall prematurely, knead oak and vine shoot ashes with vinegar, mold the mixture over the roots of the grapevine, and cover it with soil.

Grapevines with many vine shoots but little fruit benefit from extensive late prunings to strengthen them and reduce the number of branches. Remove their branches and leaves, excavate their bases, and fill the holes with river sand and ashes. It is best if their bases remain excavated all winter.

To ensure that aphids and other insects known to seriously harm grapevines do not breed in a vineyard, avoid situating it in a humid region and, many will say, rub the pruning knife repeatedly with bear grease or ground garlic dissolved in oil to maintain the odor—a method I do not consider very reliable. According to agriculturalists, fumigating the vineyard with wax and sulfur will eradicate all aphids, worms, and ants. In addition, it is recommended that the vineyard be as free of weeds as possible, since aphids grow among weeds, particularly in humid regions without much wind. It is also necessary to remove preexisting aphids from the stocks by placing bags with wide mouths and narrow bottoms over them then shaking them so the aphids fall inside. This must be done before the aphids lay their eggs, called *carrochar*, which are similar to fly eggs sometimes found on the undersides of leaves. All the leaves that have eggs must be removed, because those that remain burn and gradually destroy the stocks. Another effective technique to eradicate aphid eggs is to remove the bark from stocks they have touched in the vineyard since if the bark remains, the eggs are preserved and will grow.

An additional useful tip is to bury a ram's abdomen with its entrails, leaving a portion of it exposed aboveground either in the middle of the vineyard or in various places throughout it to attract small insects—aphids, locusts, beetles that eat tendrils, and others—that congregate on it, all of which can then be easily exterminated. For aphids, this should be done immediately after first seeing one.

If ants are present, applications of olive juice and flax water eliminates them. Also, placing a very old, twisted ram's horn in the vineyard causes them to gather on it, facilitating their mass extermination. For ants approaching vines from outside, a large circle of ashes at the base of the stock or tree thwarts their advance, but for exterminating them nothing is more effective than flax water. Moreover, fumigating any tree or stock with sulfur will also eradicate aphids or ants. It is said that placing a fig tree root, particularly that of a wild fig tree, over the grapevine or stock achieves the same result.

Also, when aphids or caterpillars are in their cocoons, it is very easy to gather them manually from young, tender vine branches and burn them. These should never be buried, however, because the earth's heat hatches them and they quickly multiply. Another type of small worm with many feet, which Pliny calls *convolvulos*, or vine inchworm, wraps itself in the leaf or tendril, gradually destroying the entire plant. It is said they can be eradicated by doing the following. Boil two large jugs of olive juice until it is as thick as honey, then add an equal portion of sulfur and boil again. Fumigate the vineyard with this mixture when there is a slight breeze, placing it where the wind originates, so it carries the odor throughout the vineyard. Fumigating for two or three days will destroy all the worms. This method is particularly useful with trussed grapevines, because the fumes rise. For grapevines attached to trees, it is often faster to douse the bases of the vines with this mixture, successfully preventing worms and other small insects from climbing them. If grapevines are old, skinny, or misshapen, it is best to tumble them, leaving only a tip exposed, although many farmers find it preferable to hoe the vines from below and graft them.

The method for tumbling, or submerging, a grapevine is as follows. Determine the direction in which the grapevine is leaning, and following it, dig a large, deep hole nearby. If the vine tumbled in that direction would obstruct a furrow or impede the growth of another stock, dig the hole in a slightly different direction. Finally, place the grapevine squarely in the hole, exposing as many tips as desired, but avoid filling the hole completely—this way the roots can grow downward.

Freezes are another common problem that causes vineyards to perish. One precautionary measure is to prune either very early, so the cuts are healthy and hard before the first freeze sets in, or sufficiently late so the cold weather is no longer a threat. In the latter case, it must be done before the vines bud. Another helpful measure is to keep the vineyard well excavated, a solution based on Theophrastus' contention that freezes tend to burn vineyards that have not been well excavated and cultivated. Just as we, influenced by sacred scripture, eliminate the thorns, brambles, and thistles of sin while tending the "vineyard of the soul," so should our vineyards be cleared of overgrowth so they can produce fruitful harvests. Certainly, everyone should endeavor to have ten well-cultivated measures of land rather than twenty that are neglected. Frequently, grapevines freeze so thoroughly that not only the new tendrils but even the stocks are burned, which becomes evident when the vines fail to grow robustly. They tend to recuperate more quickly if they are hoed from underneath.

In regions that often freeze, snow poses another problem, burning every grapevine bough it rests on. To reduce potential damage, after a snowstorm shake the snow off all the stocks. The impact of snow and ice is greatest in moist lands. For instance, vineyards located in sandy places do not freeze as much as those in dense regions, and those at higher altitudes freeze less than those situated on plains and in valleys.

Other common forms of damage to grapevines are wounds and contusions—generally caused by plows, hoes, or animal bites. If more of a particular bough is injured than healthy, remove the entire bough. If it is broken, clean the wound well and remove the damaged portion, or boil unsalted olive juice to the thickness of honey, cool it, and rub it over the wounded area, finally molding sheep or calf manure over it and covering it all with soil so the injury is well protected. All plants, especially when small, should be protected from gnawing livestock, because nothing is more damaging to vines. If animals do gnaw at a grapevine, find a healthy bud underneath, cut the vine shoot at the bud, and it will sprout again from there. The vine will grow more from this bud in a year than from the gnawed section in four years. If a bud is not to be found, hoe or graft the damaged portion and cover it with soil so it can sprout anew, as grapevines tend to grow where they have been wounded.

GUIDELINES FOR PLOWING OR EXCAVATING VINEYARDS

The recommendations for plowing fields described in chapter 1 apply as well to vineyards. The purpose of excavating, or tilling, on the other hand, is to benefit the growth of grapes by moving soil around the stocks without endangering the vines. Nothing preserves the earth's moisture better than repeated excavations, which strengthen old and especially new vineyards.

Ideally, a vineyard should be worked without the use of plows or beasts of burden, particularly if the vines hang low. Dense, rich soils should be worked while they are dry, to prevent them from hardening and becoming claylike. Working sandy areas with loose soil before a rainstorm will improve their absorption of water.

Vineyards with dense, rich soil should be excavated three times, the first one just after pruning and before the buds have sprouted, to avoid accidentally removing the young tendrils. The procedure is as follows, even for vineyards that are pruned just before winter. The area around the stocks should be excavated and all the weeds and grass removed. Because weeds provide openings in the soil that allow ice to burn the vineyard, their removal is essential to the survival of the vineyard. If the soil is thick with weeds, it should be excavated during the waning moon so they will desiccate more quickly. As for grass, it drains the vines of nutrients and should therefore be removed at least from the base of the stocks. Palladius contends that grass can be eradicated when the vineyard is excavated with a copper hoe tempered with male goat blood. Bringing pigs into the vineyard from the end of harvest until spring is highly beneficial since they not only eat weeds and uproot grass but also destroy anthills and dig large holes, further aerating the soil. Unfortunately, this approach will not work in vineyards with olive trees, because throughout the winter months olives remain on the trees, and the pigs will likely consume the entire crop.

The second of the three excavations, called *binary,* can be performed either before or after the vineyard blooms. It should, however, be done before the weed seeds mature since otherwise the tilling is apt to sow rather than destroy them. Excavators engaged in the binary phase must proceed very cautiously to avoid removing the fruit and new branches, or damaging the stocks. There is no need to hoe as deeply now as during the first excavation.

The third excavation, called *terciar,* is unnecessary in lands with loose soil that is neither moist nor weedy. During this tillage the hoe should be dragged very lightly, simply to raise the dust, which in humid places makes the grapes more flavorful and likely to ripen quickly and spoil less frequently. This excavation is better suited for humid rather than dry regions, where it may further dry the grapes, and it should take place only when the bunches are very firm, so as not to burn them and thus diminish the amount of fruit produced. Similarly, it should be implemented in mornings or afternoons, but not in the heat of the day, when tilling is apt to burn both stocks and fruit. During the summer, it is good to excavate once a month to prevent the soil from hardening and the tender vines from drying out.

In vineyards with sandy, loose soil, all the soil surrounding the stocks to be excavated should be removed and replaced with loose soil from between the rows. In humid regions, the stocks should be covered loosely, except for the roots, which should be fully buried, whereas in hot, dry regions the soil surrounding the stocks should be replaced with loose soil. Even in these areas, I oppose "earthing" the plants—that is, covering the stocks with soil—for it causes great harm. Instead, the soil should remain level with the roots, as indicated previously.

In addition to soil composition, recommended excavation methods depend on the nature of the grapes themselves. For example, tender-skinned grapes that may spoil quickly should not be repeatedly excavated, even if they are located in a humid region, and they should be excavated only deeply enough to eradicate the weeds. By contrast, firm, dry grapes like the *alvillas* are not harmed by deep and repeated excavations, and actually benefit from humidity, particularly in terms of growth, though not wine production. For all grape varieties, it can be said that the drier the soil, the better the wine.

Although digging thoroughly around grapevine stocks with a hoe, as occurs with the first two excavations, is preferable to plowing, vines trussed in trees can indeed be plowed without being endangered. It is best to plow these trussed vines as deeply as possible, while taking care never to disturb the roots or let the plow touch the grapevine stocks. And as with excavation, robust, moist soil should be plowed more times than loose soil.

Generally, the more a vineyard is excavated or plowed, the more fruit it produces and the better it will be. In humid lands, though, where too much plowing or excavation generates excessive moisture, the wine is improved when the vineyard is worked less—preferably, just to eradicate weeds and grass. One more precaution: vineyards should be neither plowed nor excavated during a freeze, as doing so can cause substantial damage to grapevine branches.

HOW TO FERTILIZE VINEYARDS

Agriculturalists concur regarding this very important advice: all animal fertilizer benefits vineyards, provided the vineyards are located in cold or humid regions; have light, loose, sandy soils; or are older than average. Fertilizer strengthens stocks so they produce more fruit. But while it is of great benefit to fruit, it is detrimental to wine production. Wine from fertilized vineyards lacks robustness, is often turbid, tastes unripened, and quickly absorbs any unpleasant flavor. Even if the wine is extremely clear, it generates sediment regardless of how many casks it is poured into, and it is has minimal longevity. Hence, to protect vineyards from becoming corrupted, farmers have sought alternative fertilizing techniques.

First and foremost, it has been found that not all lands need to be fertilized, and among those that do, few require the same quantity or type of fertilizer. Soils that are naturally dense and rich with nutrients, especially in hot regions, do not require fertilizer. Using fertilizer on these lands makes them excessively lush, causing grapes to spoil or yield an inferior wine. Fertilizing in hot, dry lands can also parch the vines, although some hot lands do require fertilizer.

In areas that benefit from fertilizer, it should come from livestock or domesticated animals, and be well aged. If it has not fermented for at least a year—or in the case of pig manure, four or five years—it is detrimental to vineyards because it produces scores of weeds and burns the stocks, while aged fertilizer is advantageous for the opposite reasons. Animal fertilizer that has aged can benefit even hot and humid soils if it is placed in a deep hole, combined with an equal amount of soil, and well covered with soil.

The fertilizing technique that is best suited to these hotter regions and other types of land prone to impacting negatively on the quality of wine, is the following. Once the grape leaves have fallen, sow lupines throughout the vineyard, or dig up the grapevines and sow lupines in their holes, or plant two lupines at the foot of each grapevine. Cover them lightly them with soil. Once they are large, tumble them and cover those too with soil. Then plow the area if possible, which will cause them to rot, adding significant nutrients to the soil. This fertilization method greatly benefits the stocks and presents no danger to the wine. Furthermore, it can be successfully performed in many different types of land: dry or humid, hot or cold, far or near.

The following fertilization technique is especially suitable for older vineyards that are trussed. Excavate the grapevines, fill a *cántaro* (pitcher) with equal parts of highly putrid human urine and water, and pour the mixture at the base of each affected vine. Ensure that they absorb it properly, then lightly cover it with soil. The mixture should be introduced during the late afternoon to prevent it from heating and evaporating in the midday sun. Grapevines fertilized with urine produce more and better fruit, and unhealthy grapevines are routinely cured of afflictions. This method is used often in Italy, where urine is gathered in the following manner. Build an underground trough with stout walls of brick and mortar, much like a vault. Build a gutter that flows to the trough in the section of the house where everyone urinates. Store the urine there for an extended period of time. This is how it is done in crowded public places that have established rooms for this purpose. In fact, the owners of these are paid well for the service.

There are yet other techniques for fertilizing. One method involves excavating the vineyard at harvest time so that the dried leaves fall in the holes and rot there. Another method, most conducive to soils that are light and lacking in nutrients, or vineyards located in hot regions, involves excavating the grapevines and integrating into the soil for each vine two basketfuls of rich soil from idle ground. Yet another possibility is to break up spoiled vine shoots and plow them into the soil throughout the vineyard. Additionally, vine shoot ash is an exceptional fertilizer when placed in holes following excavation during November or December, so the forthcoming precipitation fully incorporates them into the soil.

In general, all fertilizing should be completed before winter, except for the urine fertilization method, which should be performed in January or February, and neither before nor after. Nor should grapevines be overindulged with fertilizer by farmers hoping to avert damage in future months when it may be unavailable. Only a small amount of fertilizer should be used, because any excess burns considerably. Furthermore, not only is fertilizer best when it is most stale, but it should be applied over an extended period of time. All fertilization with livestock manure lasts for five or six years, while any of the other fertilization methods described should be performed every two years.

BRANCH REMOVAL AND VINEYARD MANAGEMENT CONCERNS

Branch removal is a rarely performed task, because few people know of the tremendous advantages it provides to both grapevines and grapes. In fact, so essential is its proper administration that branch removal skills are as valuable as pruning expertise.

Branch removal provides several benefits. It concentrates the strength distributed among young tendrils and irregular branches into fortifying the stock. As a result, the branches become thicker, and the grapes ripen better and are less prone to spoiling. Also, because the stock is cleaner, unencumbered of debris and unwanted vine shoots, the plant can be pruned easily and more effectively. And when branches are removed while still young, grapevines do not bleed, thus producing more and better grapes. In addition, branch removal rarely damages grapevines, and when it does the vine heals itself rapidly.

The removal procedure is as follows. All that sprouts between the boughs at the base of vine and underground should be removed, leaving only what is required to cover the stock, as well as any excellent vine shoots suitable for transplanting. Vine shoot tips also should be removed, to prompt strong renewed growth. Moreover, if tendrils are long enough to become entangled, tearing vine shoots from the mother plant when a person or animal trips on them, these too should be removed, to prevent them from damaging the stock.

In general, branches can be removed most efficiently while they are still tender and green, from the time of blooming until the vine shoots reach maturity. This allows for the elimination of buds that are emaciated, bent or curved, or likely to sprout in inauspicious places, such as on the stock. Removal should be done on a daily basis, because many branches tend to remain and new ones constantly sprout. To foster the growth of healthy vine shoots, remove the "grandchildren" that sprout beneath their leaves.

Later branch removal applies to grapevines that produce excessive quantities of fruit which, if left on the vine, can at once fail to mature and cause grapevines to perish—as occurs with animals when they give birth to more offspring than they can adequately nourish. Thus, if a grapevine produces so much fruit that it seems incapable of sustaining it all, evaluate the bunches and identify those most unlikely to mature, which are usually the smallest and most shriveled. These should be left on the vine for the time being, since with sufficient humidity stocks can endure an overabundance of fruit fairly well; but ultimately, should the bunches in fact fail to ripen they must be removed.

After implementing the aforementioned recommendations for branch removal, two tasks remain. If it is intensely sunny and the grapevines dry out, they can be covered with branches or leaves. If it is a humid year, or if the vineyard is located in a region where grapevines tend to spoil, the leaves should be removed, allowing the sun to dry the grapes.

Finally, I include some thoughts on overall management of vineyards. Owners of sizable vineyards will undoubtedly need employees who can help with the many chores. During one period, they can dig holes, plant, transplant, or tumble vine

shoots. At another time, they can fix fences, deepen ditches, close gates, trim trees and trusses, dig, excavate, clear briars, cut props and sticks, and elevate twigs. They can also prune and graft if they possesses these skills, and dry fruit as well. Likewise, they can guard the vineyard in winter, protecting it from livestock that might trample or gnaw the vines, and in the summer they can be of constant service guarding the fruit from livestock, especially sheep, whose wool wraps around the buds, permanently causing them to blossom poorly, and from the gnawing of goats.

While no owner can attend personally to all the chores in a large vineyard, and hiring others is costly, a well-tended vineyard will result in a good profit. What they say is undoubtedly true: when good is done, it is repaid by good.

Ways to Store Unripened Grapes

Before the harvest and the rainy season, figs, peaches, and *priscos* (peaches resembling nectarines) can be dried and sweet and firm grapes can be gathered for making raisins or storing in a not-quite-ripened

state. Any damaged grapes should be removed from the bunch, stem and all, with a small pair of scissors. One spoiled grape ruins the entire bunch just as one spoiled bunch damages all bunches in storage, due to the foul odor and microorganisms the bacteria emit.

Grapes should be gathered during the waning moon and after the sun has thoroughly dried both the morning dew and the night's moisture. If it rains before they are gathered, they should be allowed to dry well during subsequent clear days.

To select grapes for storage, look for very dry fruit that is neither too plump nor too small. Their skin should be firm, and the bunch long and sparse. The *jaén* variety of grapes is particularly well suited for storage, provided they are grown in a dry area. Those grown in sandy soils are better than those from rich lands. *Hebenes* or *palominas* are also well suited for storage since their bunches are sparse and their fruit dry and firm. To avoid becoming bruised, they should not be transported on animals from the vineyard to the storage place. Grapes maintain their beauty and intactness best when carried on the heads of maidens.

There are several methods for storing unripened grapes. One method is to hang the gathered bunches so they do not touch one another—thereby allowing for good ventilation, preferably on small rooftop platforms surrounded by grates and a net to protect them from the birds, as the Moors did in Granada. The place itself should be well ventilated too, preferably windy (since stagnant air rots grapes), cool, and without humidity. Apples, *peros* (a kind of apple), quince, or pears should not be hung near the grapes, nor should other fruit, because different varieties can damage them. Considering how watery most unripened grapes are, it is important to hang them in keeping with the foregoing specifications for making them drier, healthier, and generally better tasting. The bunches can also be hung in wheat granaries or buried among barley or clean scattered straw.

Another method agriculturalists recommend for storing grapes requires making a fire in the vineyard to melt pitch and, after cutting the bunches, placing their stems in the pitch to seal them. Next, transport them carefully to the storage place. There, using a pair of very wide identical earthen tubs, place a bed of clean straw on the bottom of one, cover it with a layer of grape bunches, ensuring they do not touch, and continue layering until the tub is full. Cover with the other tub, seal the "cask" together with clay and straw, or plaster. Alternately, bunches with skins dipped in pitch can be placed in earthen jars between layers of barley bran or very dry poplar sawdust, then sealed. The earthen jars or casks should then be transported to a cool, dry place for storage.

Yet another method involves the use of a very wide and flat earthen jar that is sealed, both internally and externally, with pitch into which grape bunches are placed without touching one another, their stems already sealed. Then the jar is covered with a clay lid large enough to fit properly over its mouth, or with an identical earthen jar, and its seams sealed, preventing so much as a drop of water from penetrating. Subsequently, the jar is submerged in a well, fountain, or cistern, with weights placed on top so nothing appears to be in the water. When the grapes are removed, they will be as fresh as if they were just cut from the vine. Grapes stored in this manner make wonderful gifts, but should be consumed the day they are removed from the jar because they quickly turn sour. Regardless of the method used to gather and store unripened fruit, the goal is to preserve it as it is naturally, so it appears to have just been picked.

There are several methods for storing unripened grapes on their vines as well. One such method, best suited for the *alvilla* variety, is the following. Craft a large earthen vessel in the shape of a bunch of grapes, with two matching handles and sealed on all sides. From top to bottom, split the vessel down the middle into two sections, then bake them and seal them again thoroughly with pitch, inside and out. When the grapes begin to ripen, leaving them on their vine insert one bunch from above into the vessel, which should be sufficiently large to prevent the bunch from touching its interior. Then fasten the vessel together by sealing the seams with clay, and using the handles, tie the vessel evenly to the grapevine so it hangs much like a bunch. Leave a small hole between the seams at the stem, ensuring that the opening does not tighten excessively. Whenever one wishes, the vessel can be opened, revealing a very fresh bunch of grapes. But they may not be very tasty.

If the objective is to keep the grapes on the vine until April or beyond, one should do the following. Find a vine shoot with numerous grape bunches that is long enough to reach the ground when curved. Dig a very deep hole at the base of the grapevine, fill it with fresh sand, then curve the vine shoot into the hole so the grapes touch neither the bottom nor the sides. Cover the hole with two or three sticks, many lily or cattail leaves, and finally a layer of soil, thus protecting the area from both sun and water. If the hole is not uncovered until April, the grapes will remain completely fresh.

A different procedure can be implemented once trussed grapevines have budded. Select a grape bunch still attached to its vine and place it inside a large broad-bottomed glass bottle with a short, fat neck, then tie the bottle safely to the truss. The grapes growing inside the bottle with the vine shoot on the exterior will be quite a sight. When the other grapes are very ripe, the vine shoot should be cut and the grapes saved for eating.

A final way to store grapes is by covering them with leaves while they are still on the vine. If they have already been removed from the vine, the bunches can be spread out under an overhang, such as on a porch where only a little sunlight can enter but no water can touch them. This method is generally used where grapes such as *alvillas* ripen late. If the vine shoot buds are cut when the grapes bloom, they grow again and tend to ripen later in the season. On occasion, they fail to ripen but can be used instead for verjuice (*agraz*), a juice expressed from unripened grapes which, while sour, is more pleasant and healthier than vinegar, another derivative of unripened grapes. To express the juice, take unripened grapes when they are very fat and sour, and pound them in a stone mortar, adding a little salt or oil for preservation and letting them remain in the sun for two or three days. Then put the verjuice in a glazed or glass cask, and store it well covered. Verjuice is considered as precious as grapes, because even in lands devoid of orange trees it will keep for a full year.

The Art of Making Raisins, Wine, and Vinegar

Unripened grapes that have been gathered and stored have a variety of uses. Most agriculturalists like to turn them into raisins, wine, or vinegar.

HOW TO MAKE RAISINS

To make *colgajos* (raisins), select bunches of the same type of grapes, preferably long, fat varieties, such as *lairenes* or others without seeds. Prepare a cauldron of clean, clear lye, made with vine shoot ashes; boil it over the fire; then add a little oil. When the mixture wastes away, add more but not too much, since dipping the grapes into very strong lye will make the raisins better and able to be stored for a longer period of time. Then dip the grapes in the lukewarm lye mixture, removing them immediately and returning them to the lye mixture repeatedly until the grapes change color slightly; dissolving ground saffron in the lye mixture will turn the raisins golden. Next, hang the grapes where they can get sun and air (but no dew or rainwater) until they are very dry. At that point, transfer them to jars or pitchers, cover very tightly, and store them in a dry place. Some farmers put a lump of quick lime in the lye mixture to make the grapes dry more quickly, but I do not believe this ingredient is healthy for the body.

A method for making raisins from grapes that have ripened on the vine is as follows. When the grapes have ripened, remove the leaves from the grapevine so that the bunches receive sun. Twist the stems of the bunches so they dry quickly, and when they are very dry, store them well. I have found it better to cut bunches from the vines at the beginning and hang them there in the sun to become raisins. Those hung in this way will not spoil if left for eight to ten days before being moved to the shade for drying. This is a good method for types of grapes with skins so thin and delicate they would be likely to dissolve in lye.

There is also a method for making raisins more quickly. When you have taken bread out of the oven and the oven is still warm, put in the selected bunches of grapes over a *corcha* (vessel of pitched cork) or a board that is not resinous, and leave them in overnight. In the morning, wet the bunches in a good *mosto* (must, or grape juice), preferably one made of white *hebenes*. Then put them in the sun to dry, and finally store them like the others. To make the raisins more sugary, wet them again in the same must and, when dry, store them. All varieties of raisins absorb sugar if you sprinkle them with a good must and allow them to dry. Before drying and storing the grapes, some farmers dip them in a must turning nearly to syrup as it boils on the fire.

To make even better raisins, hang bunches of grapes in the sun for ten to twelve days, *then* dip them in lye that is strongly colored by saffron. Grapes with the most tender skins, having been exposed to the sun, can be safely dipped in this lye because they will by then be hardened.

HOW TO PRODUCE WINE

To make excellent wine, it is best to prepare well in advance of the harvest. This means acquiring such items such as hutches, panniers, barrels, and wide-mouthed baskets. In lands where tubs are used, they should be encircled with old hoops so the wood will dry and contract in the summer, and later swell with the humidity. The tubs will have to be washed well so that afterward they will only need to be rinsed with water. In addition to preparing the many objects needed to assure perfection of the wine, the attentive farmer will also arrange to make the most of vintage time, maximize the potential of his *bodega* (wine cellar), and know exactly how to set the casks to ferment, preserve the wine, and eliminate defects that might arise.

VINTAGE TIME

In some lands vintage time arrives early, while in others it comes late. Vintage time in any particular region is determined by when the grapes have become perfectly ripe. If they are harvested while underripe, the wine will have little vigor and longevity; if they are harvested when overripe, the wine will be murky, sweet, and quick to sour. The following signs indicate grapes' readiness for harvesting: bunches that are clear, dry, blushing, with a pleasurable sweetness; seeds that have changed color inside; white grapes turning to a drab color, or dark grapes turning darker; the removal of a grape from a dense bunch causing the other grapes to fill in the space in just one day; grapes remaining the same size day after day; the squeezing of a grape causing seeds to come out clean; and bunches so heavy that they curve the vine shoots.

When grapes are flushed and well dried by the sun, the wine will be vigorous and have great longevity. If they are greenish or wet, the wine will be bad and short-lived. To prevent negative outcomes of this sort, do not gather grapes when they are wet, even if only with dew, but rather let them first dry in the sun; or after vintage time, when the wine has begun to ferment, pour it into another jar so the water can stay at the bottom of the first one, since it is heavier than wine. Regarding the grapes left to dry in the sun, if they are very hot they should cool before being processed. When the vintage takes place during the waxing of the moon, grapes make more wine; when it occurs during the waning moon, the wine has greater longevity.

Some precautions to take at vintage time for farmers wishing to avoid compromising the wine's taste and longevity are the following. Carry a knife for cutting the bunches and removing seeds, in lieu of squeezing bunches of damaged grapes when taking off stems. Further, remove leaves from the bunches, as well as unripe, spoiled, dried up, or withered grapes. Also separate the varieties and grades of grapes, reserving some casks for good grapes of each variety and others for those not so good; for example, grapes gathered from vine shoots should go in one cask, and those from twigs in another. If a large quantity of small bunches remain on the vine

after the vintage, use them to make a separate batch of wine for drinking in the winter; if there are only a few, save them for eating.

There are three ways to produce wine: it can be made in the vineyard, the grapes can be tread there and the must brought to the house, or the grapes can be brought whole to the house and the wine produced there. Of these methods, the owner of the vineyard should use the one that best suits the situation. To make wine in the vineyard, as is done in Córdova, one must have a vineyard containing a special single-story house called a *lagare,* with a wine cellar and troughs where wine is fermented, allowed to settle, then brought clean to the main house for decanting. Although the farmers of Córdova sometimes produce bad wines, this is due not to the way they make wine but rather to the incompatibility of grapevine and soil. Because the soil of the region is rich, they should plant delicate grapes such as *alvillas, cigüentes, castellanos,* and *moscatels,* but instead they plant dense varieties—and dense grapes grown in rich soils make a wine that is richer and thicker than a workhorse's urine and lacks longevity. Even so, their way of producing wine has many benefits. The *lagare* offers everything necessary to produce the wine quickly: shelter in the event of rain, a place for laborers to sleep so they can start work early and stop late, and proximity to the vineyard, resulting in little if any loss while transporting the grapes and the added boon of grape skins being left behind as fertilizer for the vineyard and food for the birds.

In Rome, wine is produced by the second method. Vineyards there contain vats made of brick or rock covered with pitch, and floors of mortar or well-layered brick. The vats are tilted slightly to one side, allowing the must to run out through a hole and into a smaller vat. There the must is collected in barrels or copper pitchers sealed inside with pitch, and transported by cart to the house, where it is put in jars to ferment, the skins having been left in the vineyard.

By contrast, here in Talavera and elsewhere, grapes are transported whole to the house, where they are put in troughs or basins, and treaded. This is a good method to use when the vineyard is within a league [2.4 to 4.6 statute miles] of the house, but when it is farther away, much must is lost, which can be costly.

Regarding the treading of the grapes, it requires skill and should be done by someone strong who can squeeze the fruit well with their feet. A bar or rope should be placed across the top of the tub so the peson does not fall. And of course, all such workers should be clean, because cleanliness is one of the most important attributes required for making wine.

When it comes time for the fermenting, there are two techniques. The first technique involves fermenting the must free of grape skins, which produces a wine as clear as water and with great longevity. So that it settles well and remains pure, skins can later be put back in to cover the stopcock hole through which the wine will be decanted. It is also possible to raise the stopcock instead, so the sediment does not pour out when decanting. The second technique entails fermenting the must with skins, which will make the wine ruddier, golden colored, and more rigorous and short-lived. Wine that is fermented in this way should be decanted once it has settled, so it will not take on the flavor of the skins. The longer the wine remains over the sediment after settling, the more short-lived it will be, unless it was made from bad grapes. Wine that has remained over the sediment after settling can be improved by passing it through the skins of better grapes then allowing it to stand for several days.

Crecentino disapproves of the way farmers in some lands tread on the grapes and put them to ferment together with their stems. In Santa María del Campo, I once saw a house with troughs in which they had put all the ripened grapes from two or three vineyards, together with their stems, leaves, and sour unripened grapes. Wine made in this way is murky, and the bad mixture with its offensive odor is better poured into the street than drunk. Wine made from good grapes, as opposed to bad or green grapes, ferments later and settles less quickly, because it has more vigor. Further, many people put the jars or casks of fermenting wine on porches where it does not cure because it receives sun or evening dew. Also remember that wine being fermented is hot and easily takes on any foul odor, which intensifies over time. For best results, the place where wine is fermented should be cold, not humid, and free of bad odors—for example, far from stables. In some areas, when treading, they put ground gypsum into the wine, which makes it whiter and slower to turn to acid, but this ingredient is detrimental to health, burning the liver, hindering respiration, inflaming the stomach, and causing kidney and bladder stones.

THE *BODEGA* (WINE CELLAR)

Ultimately, one of the most important things for the preservation of good wine is the place in which it is stored. If wine is not stored in the proper manner, care in wine production will be of little value. As with a married couple, if the husband earns and the woman saves and spends wisely, they become rich, but if he works and the woman throws the money away, the estate will not flourish. Similarly, a good vintage stored in a good wine cellar will rarely result in bad wine, but if the wine cellar is bad it may adversely affect a good wine. And if a bad vintage is stored in a bad wine cellar it will be astonishing if there is one good drop.

There are two general types of wine cellars, one underground and the other above ground. The best type of underground cellar is one that has been dug into solid rock. There are some like this in Sutria, close to Rome, in a place called Las Ferreras, near Susa, and in many other areas. These are excellent because in the summer they keep the wine very cold, and they are warm in the winter. Others are dug underground, where there is argil or rich clay, such as those found in Campos and in Alcarria. While these are also very cold, they are commonly humid, which is undesirable for wine. Still others are natural underground caverns, or caverns made of wood, the former being better. Another type of underground wine cellar called a *soterraño*, involves much work because here the casks are first buried then unearthed. They should be buried in sandy soil, as it is cold. It is also necessary to be sure they are very tightly covered so small insects—especially cockroaches, which are filthy and have a great stench—are not likely to fall into the wine.

Wine cellars above ground, on the other hand, should always be very cold, closed, and in a place without wind, unless it is north wind. The floor is best made of mortar, or mortar and brick, and built lower toward the center, site of a small basin of rock or clay designed to collect water and, in the event that a cask spills over, some wine as well. For this reason, the floor should be routinely cleaned.

To be effective, every wine cellar no matter how it is made, should have the following characteristics. It should be deep, cold, dry, and dark, with thick walls and a very strong roof. It should be far from baths, stables, rubbish heaps, sewers, wells, smoke, vinegar, granaries, and trees, especially fig trees or male wild fig trees because they harbor mosquitoes. Since wine is hot and porous, it quickly takes on any good or bad odor, especially while fermenting; hence the wine cellar's windows should face the north wind, be sheltered from the easterly wind, and be able to close tightly. In the summer, windows should be shut during the day, especially to keep out the easterly wind or any warm wind, and left open at night. It is likewise good to have the wine cellar divided into sections: one for fermenting the wine, another for keeping the clear wine, a third for settling the wine, and a fourth for storing the wine.

THE FERMENTATION PROCESS

The casks for fermenting wine, and also for preserving it, are of two types, each with features that affect the wine. Some (tubs) are made of wood, while others (jars) are made of clay. Wine is more fragrant when it is fermented in tubs, as opposed to earthenware jars, though in jars it is less murky. The jars are also less costly and dangerous than tubs, which each year require new hoops; pitch, since the hoops can easily break or leak; very strong, thick hemp ropes, in the event that a cask must be tied after breakage; and billy goat grease for sealing junctures. Considering the upkeep and expense required by tubs, jars are a most suitable alternative and, once sealed with pitch, can last up to twenty years. To do so, however, they should be made of a sandy clay free of strong odors and well baked so they are both light to carry and unlikely to absorb the wine.

Regardless of their composition, casks for fermenting should be large, wide mouthed, and big bellied, whereas those for storing the wine should be of medium size, round, and narrow mouthed with a long neck. It is better to have many medium-size casks than a few large ones because the larger quantities of wine can spoil before it is all drunk.

Casks for fermenting should also first be cured. Tubs, if they are old, must be scraped well inside to remove any dirt, mold, or pitch, then kept in the sun for a day before being encircled with hoops. To prepare to cure them, collect long, thin bundles of very dry branches, preferably from vine shoots, even if they are covered with pitch. Set one bundle on fire and, holding it, use it to warm a tub's exterior while two other workers roll the tub around, tipping it from one side to another until the heat can be felt on all outer surfaces. Then apply melted pitch, rolling the tub once again until the pitch congeals and spreads over the entire surface area. (To keep the taste of pitch out of the wine and simultaneously remove any mold, two liters [one and a half quarts] of hot vinegar, in which a good quantity of salt has been dissolved, may be added to the melted pitch prior to application. Also adding finely ground cumin and some cloves and ginger to the melted pitch gives the wine an exquisite taste and helps preserve it.) Next, spread pulverized pitch around the mouth and other parts the melted pitch could not reach. Once the pulverized pitch has congealed somewhat, pour a cauldronful of lukewarm water into the tub, tip it back and forth until the pitch is fully congealed, then remove the water. Follow by filling the tub with clean, cold water, which will help it take on a shine. If tubs are new and of good wood, they do not need to be cured.

Earthen jars are cured differently. First, keep them in the sun for a few days. Then on a calm day, turn them upside down over rocks and expose them to fire until their bottoms become too hot to the touch. Now melt a little pitch, preferably mixed with vinegar and salt, and apply it to the inside bottoms of the jars; alternately, pieces of pulverized pitch can be applied, although this method is less effective. Next, reaching inside each jar, seal its mouth and neck with pulverized pitch that is neither smoky nor dirty. Good pulverized pitch is transparent when you break it, has a golden hue, and tastes sweet, whereas smoky or dirty pitch is bitter and harms the wine no matter how much it has been dressed with vinegar. Iris roots, dried and ground, may be added to the pitch, helping the wine take on a good fragrance. It is also possible to mix wax with the pitch to prevent it from splattering, although Pliny says wax deteriorates wine.

Casks holding wine should be as clean as the glasses from which it is later drunk. Consequently, both types of casks should be washed repeatedly until the water comes out clear. This holds true not only when they are being cured but also after they are decantered or later emptied of wine, since sediment from wine and dirt, if left inside, gives the next batch of wine an unpleasant aftertaste. Further, agriculturalists say that if vineyards are near the ocean, casks being cured should be washed with clean ocean water, because its salt content will remove mold and preserve the empty casks. Another option is to wash them with ocean water that has been mixed with samphire, pennyroyal, anise, or any other good fragrance and boiled. Or one can hang a perfuming pan inside with burning embers and incense, then cover the cask so it receives the smoke. After the casks are cleaned in one of these ways, the wine can be poured into them, never filling the casks to the top or they may spill over. Just in case, for each type of wine take along a small cask that can be used to catch the excess.

The best way to set the casks to ferment wine is as follows. In advance, around the walls of the wine cellar make a bench of brick or stone about the height of two palms and wide enough to hold the jars or tubs and still be able to walk among them. Pliny recommends separating the casks so that if one is harmed it does not touch another. The stopcocks should be low enough to alleviate the need for tipping the tasks for decanting. Those containing white wine should be set apart from those with red wine, and the best wine should be set apart from the less good wine, with one stirrer for the most desirable wine and another for the rest—or if only one stirrer is available, plan on using it first for the white or the best wine then washing it before using it for the other wines. The purpose of stirring is to disperse the vapors of the wine and prevent it from souring, or otherwise turning bad. Before stirring, however, always check to see if a small insect or dirty or poisonous object has fallen into the cask and also remove anything near the brim. As a precautionary measure, keep each cask covered with a cloth, or a hemp sieve.

When wine is fermenting, any exquisite fragrance can be given to it. Some people break up cloves, hanging them by a string in a clean, small linen cloth inside the cask, while others add ginger, cinnamon, or musk. Such fragrances significantly protect the wine from taking up bad flavors. Also at this time, white wine can be made dark, or into a mixture of white and dark, by passing it over the skins of dark grapes. And, according to Aristotle, sweet wine can be made by adding oregano. Or raisins can be added, a few at a time, to enhance the wine's fragrance and longevity. Or sweet wine can be made of raisins alone, by pounding them a little until they break open then putting them in a cask and fermenting them with water. When they settle, they may be removed through the stopcock or with a clean lock of wool used as a filter.

Technically, both before and during fermentation grape juice is known not as wine but as must, a beverage in its own right. To store must so that after a year it tastes as fresh as when it was squeezed, Columella recommends taking the first must that comes from the grapes, straining it well, and putting it in a pitcher or jar that is then sealed thoroughly and set in a well or cistern with sweet water. It will keep for a year and, when subsequently drunk, will cure digestion and repletion problems in children.

FROM FERMENTING TO PRESERVING

After fermenting in casks in the wine cellar, wine that is to be stored there must first be decanted then put in clean casks, dressed as necessary, and sealed well. The decanting—the passing of wine from one cask to another—should be done in cold weather or when there is a north wind or on a clear and calm day, since rain or even cloud cover discourages wine from slipping easily out of the grape skins. Crecentino further advises against decanting when vineyards are budding and rosebushes are beginning to bloom, explaining that at such times wine's sediment is disturbed more than ever. To decant the wine, pour it, sediment and all, from the large cask into a medium-size cask for storage and purification. Wine from fertilized vineyards should be decanted more than once because of the great amount of sediment it makes.

Once the wine is in its new cask and settled, mature orange peels, the skins of *peros de nelda* (a variety of pears), orange blossoms, or pippins can be added to both preserve the wine and provide an exquisite fragrance. After a while, however, the wine will need to be strained into another cask for storage, since fruits that ferment for too long can cause the wine to take on some of the bitterness produced by their fermentation. Wine that has been strained can be put in with other wine, as long as the skins are omitted.

People who wish to store wine for some time should remove the wine at the top and bottom of the cask, saving only the portion in the middle, which has the most strength. Hesiod and even Macrobio assert that the best wine is in the middle. Often we see that in casks touched by the easterly wind, only the wine at the top is harmed, but more often the easterly wind causes wine throughout the cask to turn sour. And while some say that wine can revert to its original state after turning sour, this is false, for by then the mixture has lost the qualities of wine and taken on the qualities of vinegar. Galen, in the *Gloss of the Aphorisms of Hippocrates*, says that it is impossible to make wine from vinegar because the latter has lost the strength of soul and spirit, which medical doctors call "quintessence."

Even so, it is possible to make wine less sour by doing the following. Add to the cask a *celemin* [about a peck, or two gallons] of walnuts hot from the oven, and cover the walnuts with cut green willow branches stripped of their bark. Leave the walnuts and branches in place for three or four days to prevent acidity.

Other additions to the cask are also possible. Seedless raisins that have been rolled wet in clean sand give wine a pleasing taste, especially if they are made from a fine variety of fragrant grapes. Or for sweetness and longevity, one can boil the must in a new clay cask until one-third of it has evaporated, then add ground or broken-up spices to the remaining foam and, when this mixture cools off, pour it into the wine.

Before sealing, make sure the cask is filled to the neck, with no breathing hole, as casks of wine that are less full turn sour more quickly than those that are full. For an added precaution against a wine's souring do the following. Take a large piece of very good, fat salt pork without meat; wash it well to remove the salt; hang it by a cord inside the cask, toward the center of the wine; then close the cask. Wine should be removed from the stopcock, since if it is taken out through the mouth of the cask the wine will be harmed more quickly. As the wine is used, the pork fat will be lowered. When all casks in the wine cellar are filled to capacity and otherwise protected against souring, their brims should be crowned with pennyroyal or oregano to ward off bad odors, after which they should be sealed with a thick film of pulverized pitch. If the sealer does not include ground cloves and cumin, or other spices of good fragrance and flavor, add them just before the pitch. During times of preserving, as with fermenting, the wine cellar should be sheltered and hot in the winter and cold in the summer.

REMEDIES FOR DEFECTS IN WINE

Some defects in wine can be remedied, while others can be prevented. One common defect that can be assessed and eradicated is water content in the wine. According to Crecentino, there are many ways to ascertain whether wine contains water. One way is by observing what happens when raw cut-up and peeled pears or mulberries are placed in the wine; if they sink, the wine has water. Another way is to take a stalk of *carrizos* [ordinary reed grasses] that grow along riverbanks or a smooth stalk from oats or rushes, smear it with oil or grease, and put it in the cask with the wine; if there is water content, drops of water will stick to the reed, and in fact, the more water there is, the larger the drops will be. A third way is to pour urine into a new earthen vessel, add some wine, then hang the vessel indoors for two days; if the wine contains water, beads of water will sweat out through the pores of the clay pot. Still another method is to wet a brick of quicklime with wine; if it has water, it will scatter. Similarly, heating oil and wine in a frying pan will provide the needed information; if the wine has water, it will splatter. Some farmers take the wine in their palm and rub it; if it becomes sticky, it is considered pure, and if it does not it contains water. Others grease a new sponge with oil, use it to cover the mouth of a suspicious cask, turn the cask upside down in such a manner that nothing can escape, then examine the sponge for signs of strained water.

For wine that is found to contain water, one of two methods can be used to separate the water. The first method involves dissolving a little mineral salt in the wine then taking a new sponge, soaking it in oil, and covering the mouth of the cask with it. When the cask is then turned upside down, water will come out, leaving the purified wine. Cato and Pliny describe the second method as follows. Make a "glass" out of dried ivy, and pour the wine into it. The wine will drip out, leaving the water; if there is no water content, the wine will drip out leaving the glass empty.

Many other flaws can also be uncovered and corrected. For example, a sign of good wine is when the pulverized pitch placed on top melts to the bottom of the cask if the wine lasts until the summer. But if the pitch is still whole at that time, the farmer should dispose of it because it is sapping vigor from the wine. Another good sign is wine that does not smell of mold. Instead, it has formed a thick skin called *flor* on the surface of the wine, which is the same color as the wine. But when white wine has a somewhat reddish skin, or the dark wine's is whitish, the wine is rightfully deemed defective and unfit for use.

The quality of the sediment, or the dregs, also reflects the quality of the wine and can be tested as follows. In the lid, make a hole small enough to accommodate a thin reed that is long enough to reach the bottom of the cask. If the open end of the inserted reed is covered with a thumb to prevent air from entering and then sucked, wine begins to come out. Covering the end and then sucking again will bring the sediment out, at which point it can be evaluated for fragrance and flavor. If it turns out the sediment smells and tastes bad, this means the wine has not been properly preserved and should be discarded. Another testing method is to bring some of the wine in the cask to a boil in a new clay cask, let it cool overnight in a pitcher, and the next morning taste it to see if it seems good or sour.

To prevent the skin that forms on top of wine from absorbing so many nutrients that it sinks to the bottom of the cask, harming the wine, remove it once a month during the winter with a small sieve. The more skin there is, the more often it should be removed. Columella says that every time the skin is removed, the mouth of the cask should be rubbed with pine nuts and covered so the wine will not be ruined.

Another major defect of wine that can sometimes be remedied is unpleasant taste originating in grapes from a bad harvest or bad land. To counteract this problem, take the dregs of good wine, form them into little cakes, dry them in the sun, then grind them up, mix them into the defective wine, and cover the cask. Another remedy is to take barley flour, make a *talvina* [a grain-based milk that, well beaten with wine, is used to make porridge and dumplings], mix it into the wine, and after letting it settle for two or three days, pour the wine into a clean cask. Still another remedy is to pass the defective wine over skins from good wine.

All wine from sandy lands that has been permeated by the easterly wind is improved by walking with the wine on sunny days, carrying it in good leather bags, then putting it in a cold place where it can lose its bad flavor. Effects of the easterly wind should be removed from wine because beyond bad taste, they contribute to headaches. Another means for eradicating effects of the easterly wind is to seal a new clay bottle and hang it, empty, by a cord in the cask of wine for a day. It is said that the bad odor disappears because the bottle collects all the contamination from the wine.

To remedy a bad odor in sweet wine that has turned sour, pass it over fragrant grape skins as many times as is necessary. Others say to take a good quality of very hot millet and submerge it in the wine as many times as it takes to draw out the sour flavor. It is also beneficial to rub sage around the mouth of the cask without letting it touch the wine, and then cover the cask. This should be done repeatedly in the morning, afternoon, and evening. An additional method is to hang a small bag with a little salt, gypsum, and dried fennel or anise inside the wine.

Mold is a serious defect. To remove it from the casks, it is good to cure them again when they are empty, or to pour good vinegar into the casks and leave it for several days. Further, a remedy for making murky wine clear is the following. Take one egg for each *arroba* [a liquid measure equivalent to about four gallons] of wine in the cask; for example, a cask containing twenty *arrobas* calls for twenty eggs. Also take a large glazed earthen tub filled with river sand that has been washed. If the wine is dark, break the whole eggs into the sand, but if it is white wine use only the egg whites, because the yolks will redden the wine. Yolks, because they spoil quickly, also make wine smell if they are left in it for a long time. Next, move the wine around with a stirrer, remove the resulting foam, stir in the egg mixture, and cover the cask. After a day and night, the wine will be clearer. In lieu of sand, some farmers use pulverized pitch instead, which is even better. Further, to minimize the bad flavor and harmful qualities of new wine that has not been cured, it is helpful to take an *azumbre* [a measure equivalent to about two liters, or one and a half quarts] of wine and boil it two or three times in a clay pot that has been rinsed well with water. Then pour it into one half an *arroba* of new wine, let it sit overnight, and in the morning notice the good flavor. It is also beneficial to melt a good quantity of pitch and, while it is hot, pour it into the wine. Other farmers water it down with boiling water then let it cool and sit.

Despite these and other imperfections, wine is naturally good, and it is better to leave it alone as much as possible rather than embark on procedures likely to limit its goodness and longevity. And while remedies may help, they rarely result in perfect wine. People are correct when they say "*De vino, ni de Moro no hagas tesoro*" ("Do not become attached to earthy possessions," or literally, "Do not make a treasure of wine, nor Moor").

HOW TO PREPARE VINEGAR

Vinegar is made from wine, and from other ingredients as well. Vinegar produced from wine occurs automatically when the wine is corrupted or more intentionally through the following methods. First, wine can be put in the sun or in another hot place, such as near a fire, in a cask that has previously held vinegar. Second, wine can be passed over sour grape skins. Third, wine, water, or vinegar can be poured into the skins and left for a few days. Fourth, vinegar can be made quickly by heating steel rods and inserting them repeatedly into the wine, each time covering the cask so the heat does not escape; this method, however, produces unhealthy vinegar with a bad taste. I believe it is better to place new, clean tiles, bricks, or smooth stones in the fire until they are white hot, then set them in the wine and cover the cask so the heat does not escape. Further, a well-covered pitcher of good wine can be placed in a cauldron of water, which is then put over the fire to boil.

Travelers and sailors who like to carry vinegar in a bag can make it in various other ways, depending on where they are. They can produce a strong vinegar by pounding together red cherries and mulberries, very sour, fat, unripened grapes, unripened mountain sloes, and some vinegar, then shaping the mixture into little mounds and setting them in wine to make them sour. Another option is to mix the above sour ingredients and, in lieu of the wine, use very sour orange or lime juice. Similarly, it is possible to pound sour sorrel, make it into little loaves, and when they are dry, add them to a crock of wine, which will suddenly turn to vinegar.

Pink- or scarlet-grain vinegars, known for their excellent fragrance, are made in the following manner. To make pink-grain vinegar, add to each *arroba* of good white vinegar one pound of fresh or dry rosebuds. (Red roses are better than white ones, which tend to turn the vinegar dark.) Pour the mixture into a glazed cask and leave it in the sun, well covered, for forty days. Then separate the vinegar from the sediment of the grain; put it in another glazed clay bottle; and keep it in a cold place.

To make vinegar of scarlet grain, put an ounce of scarlet-grain powder, with which fine cloth is dyed, into a *media azumbre* [about a liter] of vinegar. Leave it in the sun, well covered, for several days. This is considered the best vinegar of all, because the grain is very aromatic. It is especially good for smelling in times of pestilence and whenever the air is polluted.

The same method can be used to make vinegar from elder-tree flowers. In fact, vinegar can be made in this way using any fragrant substance.

Medicinal Properties of Grapevines, Raisins, Wine, and Vinegar

ABOUT GRAPEVINE CURES

Grape leaves can be used medicinally to cure numerous ailments. When boiled in water and placed on the body where the pulse can be felt, they cool fevers. Green grape leaves pounded and placed over wounds, clean and heal them. When placed on the head, they cure headaches and cause drowsiness. Drinking the water from grape leaves breaks up kidney stones and thins facial skin stones; and washing the eyes with it clears the vision. Further, ash from trussed vines, when made into a plaster with rue juice and oil, is good for poisonous bites. In fact, fire from vine shoots is healthier than fire from any other wood, and it is long lasting, provided the vines are cut during the waning of the moon and waning of the day.

As for the grapes themselves, the fresher and moister they are, the more they take on the properties of phlegm, causing inflammation, sore throats, and toothaches. Because of this, the *alvillas* are best. The *moscateles* cause headaches, but this tendency can be minimized by keeping the grapes in cold water for a while or gathering them in the coolness of morning. Bunches that have been hung are better, because they are not as moist. To be most easily digested, grapes should be eaten at the beginning of a meal. At vintage time, and other times when there are fresh grapes, people eating many of them should also have sour pomegranates to reduce the accompanying phlegm. To counteract any harmfulness of fresh grapes, physicians advise eating them with walnuts, almonds, or hazelnuts. That is why, on the tables of grandees, whiteners are served in rose or orange flower water.

In terms of topical applications, sour grape skins placed over a fresh burn prevents blistering. It is also beneficial to cover burns with syrup made from grape juice, rinsing them many times with cold water.

ABOUT RAISIN CURES

Raisins soothe the stomach, clear the voice, clean the lungs, and are nutritious, the black ones made from dark *castellana* grapes being the best. Those without seeds are good at the beginning of a meal, while seeded raisins are good after a meal—in either case, they make whoever eats them very happy, says Magnino Milanes. When eaten during fasts, they clear the liver, improve a bad constitution, and aid the memory.

Raisins are also medicinal when dressed in any of the following ways. After the grapes are cured, each bunch is sprinkled with a white, aromatic wine; coated with ground cinnamon, sugar, and cloves; then placed in an earthen jar in a dry location. Finally, each bunch is wrapped in its own leaves, with citron and laurel leaves inserted throughout. Citron and laurel leaves can also be placed between raisins to help preserve them and give off an excellent fragrance. Raisins prepared in this manner soothe the stomach. If they are instead cured in wine, they help fight coughs caused by colds. By whichever method raisins are made, they should have laurel, orange, lime, citron, or any other very hot leaves mixed in, provided they are not rosebush leaves. Dressed or undressed, raisins benefit the body in many other ways as well: they help the memory, aid in sleeping, assist in digestion, and are very nutritious.

ABOUT WINE CURES

Pliny calls wine the blood of the land. If drunk in moderation, wine is healthy, but when drunk to excess it is harmful. It burns the blood, lessens inner and outer strength, increases anger, weakens the nerves, and diminishes natural body heat and even ingenuity. It takes away memory, clouds understanding, numbs the tongue, shortens one's days, sows much discord, and causes people to reveal secrets, waste estates, and dishonor others.

Further, excess wine is known to cause conditions such as gout, paralysis, itching, and leprosy. That is why Plato made laws ordering soldiers not to drink, especially to excess. Soldiers need to be vigilant, as Titus Livy says of Hannibal. Moreover, since wine burns the flower containing seeds of fertility, couples who drink too much beget children who have diminished strength, ingenuity, and abilities. Because wine diminishes strength and sanctity, the angel ordered Samson's mother not to drink wine before she became pregnant. And the mother of the priest Samuel clearly told Hely she had refused it.

Further, Solomon and Daniel relinquished wine to be more enlightened in their knowledge of God. Even the glorious doctor Saint Domingo abstained from wine for ten years to better comprehend divine and human letters. Many gentiles also drank little or no wine, to gain clarity. When asked how he had become so eloquent, Demosthenes, the clearest orator in ancient Greece, responded that he had spent more on oil for his lamp than on wine for drinking. If it is so harmful to common people, it must be more so to those in charge of governing and justice, as God protects princes and others who govern.

With regard to the drinking of wine, fragrant wines cause inebriation most quickly and headaches as well, because the humors rush to the head. Those who drink dense, dark wines, especially if they exercise, retain heat in their kidneys, form stones, and have difficulty urinating. Each person should therefore use restraint and moderation, or take wine solely for medicinal purposes (as Saint Paul, the apostle, advises Timothy). To avoid getting drunk, it is recommended that people eat roasted sheep lungs before a meal that is to include wine. Commenting on this idea, Pliny writes, "Judging by how much is drunk in our time, I am certain that an oxen's lung is not sufficient, much less that of a sheep." There was never a drunkard who lived healthy, or who died at an old age. Another way to avoid drinking wine is by eating raw herbs with vinegar. Yet a further way to discourage wine drinking, according to Pliny, is to put two or three large eels, or dried fish, into a pitcher of pure wine and give it to those who like to get drunk. If they eat the eels while drinking the wine, they will have a great dislike for wine after that.

Regarding the use of wine for medicinal purposes, dark wine is good for people who are dehydrated, such as cholerics and even bleeders. It is also the best type of wine for people with gout if they must drink. This is because it is more restrictive than the others and does not allow humors to flow to members of the body. White wine is good for sweaty people, such as those suffering from inflammation or afflicted with stones. Both white and red wines are for people with melancholy.

What in France is called *clarete*, a wine midway between white and red, is good for all constitutions.

ABOUT VINEGAR CURES

Vinegar has qualities that make it useful for numerous medicinal purposes. Vinegar is in general cold and dry. White vinegar, which is superior to dark vinegar, is penetrating, cutting, and constricting. Placed over inflammations, it constricts the pus inside and reduces swelling. It prevents vomiting when placed on the nostrils and is thus useful for sailors, travelers, and others who have difficulty keeping food down. Foods that contain vinegar, such as parsley sauce, not only improve the appetite and aid digestion but also refresh the liver. Rubbing vinegar on the body where the pulse can be felt reduces the heat of tertian fevers, while washing the body with it renews the skin, washing the mouth with it tightens the gums, and gargling with it cures sore throats. Drinking it hot is good for people who have food poisoning. Eating soups that contain it quenches thirst and also kills worms and hiccups. Bathing with it soothes sunburn. Washing the head with it prevents hair loss and dandruff. Washing with it repeatedly will stop itching, and applying it to the skin combats ringworm. Because it tends to keep the blood from coagulating, it is good for treatment of *landres* (a morbid swelling of the glands), and draining infections.

Moreover, vinegar is medicinal when placed on the bites of dogs, scorpions, spiders, centipedes, poisonous snakes, and many other venomous animals. Pliny tells a tale of a young man who carried a skin of vinegar and was bitten by an asp, after which, every time he stopped to rest and put the skin on the ground, he felt more pain than when he was carrying it. Noticing this, his companion felt that the vinegar was good for combating the poison, so he gave him some to drink and washed the wound with it many times. It is even said that vinegar relieves foot pain due to gout, and hot vinegar soaks heal the feet. Further, an excellent cure for measles is to wet a cloth with vinegar and place it on the rash. When making a poultice, it is best to mix vinegar with other juices or water, because when used alone, vinegar chills very quickly and can be harmful. Its potency increases with one's use of salt, pepper, or other spices. And although negatively affected by the easterly wind, this impact dissipates over time.

Vinegar is so strong that it can be used to break apart boulders, something Hannibal resorted to so his provisions could pass through very rocky regions en route to fighting the Romans. Anyone can do this by first making a large fire on top of the selected boulder. When it is very hot, pour strong vinegar on top, which will cause the boulder to crumble like dirt. Vinegar can likewise harm the bladder and the nerves. Paralytics, people with gout, and those with pain in their sides should stay away from vinegar. Its use is indicated for cholerics, but melancholics should avoid it. Vinegar accentuates the aging process, making people who use it grow old before their time. Vinegar is beneficial when inhaled by women who have difficulty breathing due to passion.

Moreover, a precious stone, whether it be a pearl, *balax* [a gem similar to a ruby], or emerald, is damaged if put into vinegar—a fact illustrated by the story of the feast the Queen of Egypt, Cleopatra, had for the Roman Captain, Mark Antony. At the end of the sumptuous dinner, to make it a better feast for Mark Antony, she removed from her breasts a *balax* that had been presented to her and put it into a glass of very strong vinegar that he had ordered brought to him. The liquid stripped the stone of its spirit and soul. Having done this *magnificencia* (the name I give to the craziness of the gentility), she reached for another, more precious stone to repeat the act, but Mark Antony forbade it. Vinegar also preserves and tenderizes meat or fish when used in a dressing along with garlic, oregano, salt, and water. Finally, Pliny says that vinegar can be used to combat a very dangerous storm at sea that sailors commonly call *refriega* [translated literally as the fray or a scuffle]. When such a storm begins, if strong vinegar is quickly thrown into it, the storm will dissipate.

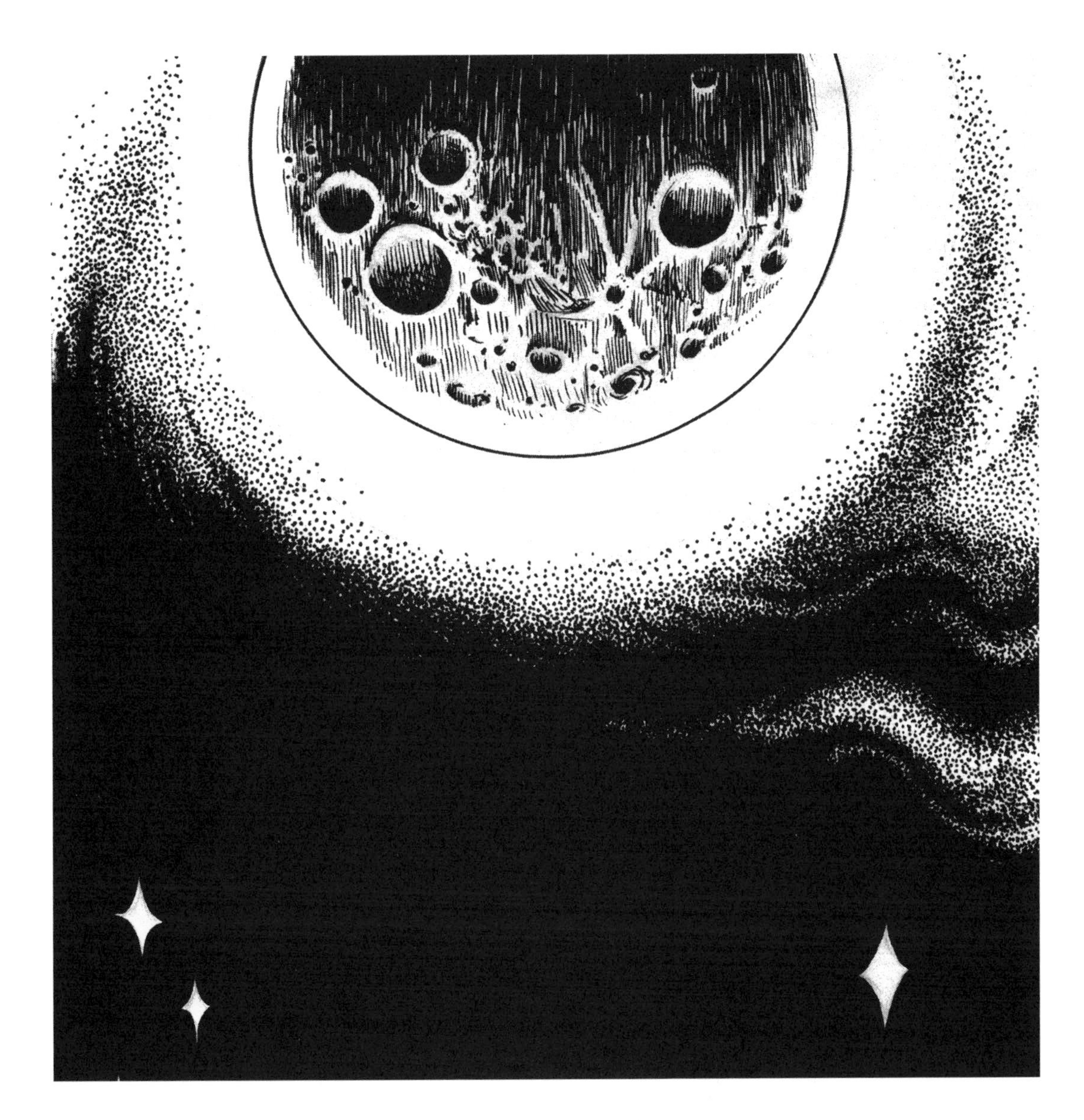

CHAPTER THREE

USING TIMING METHODS BASED ON ASTROLOGICAL INFLUENCES

Times for performing agricultural tasks appropriate to moon cycles and weather indicators

THE PURPOSE OF THIS CHAPTER is to provide clear reminders of appropriate times and conditions for performing certain agricultural tasks. Because not everyone understands the intricacies of astrology, I will offer advice that is easily comprehensible, regardless of a person's knowledge of this divine science. In fact, many who boast of astrological expertise may actually know very little about the subject, and even those who know quite a bit are not always correct. In the words of Saint Matthew, the Lord asserts regardless of visible celestial tones and signs, suggesting that we will never completely understand the complexities of upcoming seasons. Of course, those who practice the science of astrology possess greater knowledge than those who do not, and such people who study the movements of the stars and planets, as well as their differing regional effects, can generally predict dry and humid years. Such knowledge leads to sowing in meadows, lowlands, and loose, sandy soils during dry years and in higher altitudes and richer soils during humid years, resulting in productive outcomes.

It is important to recognize that while the signs I will provide are quite clear, they may at times prove to be incorrect since here on earth our inherent deficiencies prevent us from fully comprehending the impact of celestial occurrences. Only God possesses infallible knowledge, as an integral part of his absolute power. Humans must nevertheless endeavor to understand such information as set forth by wise, highly esteemed individuals, since it generally proves reliable. [*Compiler's Note:* The timing for planting each crop is based on the climate of southern Spain, which is similar to that of zones 7 and 8 on the United States Department of Agriculture Plant Hardiness Zone Map. Readers are encouraged to make the necessary adjustments to their own zone designation or its equivalent.]

The Waxing and Waning Moon Cycles

This section specifies agricultural tasks to be performed during the waxing and waning phases of each month's moon, starting in January. It should be noted that the tasks suggested for a moon's waxing phase can also be performed during its opposite phase, and vice versa, but outcomes may be less successful.

JANUARY'S WAXING MOON

Vine shoots should be planted during this month's waxing moon in warm, early-harvest regions, preferably by burying the grapevines, a practice referred to as "tumbling their heads." In early-harvest regions vine shoot scions may be transplanted as well. Early-blossoming trees, such as almond, also should be transplanted, provided they are removed from the mother tree during the waning of the day, or late afternoon. Ideally, plants should be removed and transplanted toward the end of the waning moon cycle so the rooting process is stimulated during the entire waxing moon cycle.

Sowing *tremesino* ("three month") seeds is preferable during the waxing moon, particularly in hot and early-harvest regions, leaving broad beans (*fabas*) for summer sowing. Chickpeas (garbanzo beans) and other legumes consumed during the Lenten season should be sown now and again later. I suggest sowing broad beans and resowing chickpeas at the end of September. Peas should be planted during January's waxing moon in dry, thin soils. Fenugreek should be sown now too, in well-plowed soil no deeper than four finger lengths; otherwise, it fails to sprout properly.

In early-harvest regions, one should transplant myrtle, laurel, and date seedlings during this time, and in hot regions, cypresses and other such trees. In early-harvest regions, young olive trees, willows, white poplars, and hazelnut trees or their seedlings are transplanted during January's waxing moon, though it is best to plant olive trees in October or November, allowing for sufficient water absorption. These trees should grow no higher than their planted depth, so the underground growth can be revitalized, generating new roots. After three years, the roots should be hoed, leaving them exposed to promote rapid, robust growth. Although vegetables usually flourish when transplanted during this month, plants can be successfully transplanted in almost any month.

Hay fields located in cold areas should be fertilized during the waxing moon, while in warmer regions they should be fertilized earlier. If the soil lacks density, it should be immediately protected from trampling and grazing. Hay fields are best fertilized during the waxing moon because the moon stimulates growth. Very fresh horse manure is the recommended fertilizer for these fields, since it consumes the grain and spurs hay growth.

Almond, peach, nut, plum, apricot, carob, and other pitted fruit trees can be grafted at this time. Planting pitted fruit trees during January is recommended in cold and humid areas, and during October or November in hot and dry regions. This holds true for most trees grown from seeds, including those from sour seeds, such as oranges, limes, and citrons, if planted in hot regions; in temperate lands, this is preferably done in April. Young almond trees can be successfully grafted into peach trees, and clingstone peach trees and apricot trees can be grafted into plum trees, using the *de coronilla* ("bark graft") technique. All grafts made in these and other dry, bronze-colored trees are more successful using this method. Plum, peach, or other seeded or pitted fruit trees grafted into almond trees produce fruit with an almond pit, provided they are grafted into the trunk of the tree using the *de mesa* ("bench graft") method and split using the *de junta* ("side graft") or *de pasado* ("bud graft") technique. Fruit trees grafted into hazelnut trees produce the same type of seed. During the waxing moon, gum trees can also be successfully grafted. In early-harvest regions, cherry trees can be grafted as well, since they bud early. In addition, this is the appropriate time to plant cane, particularly if the weather is humid. Seeds of orange, citron, and similar trees produced the previous year should likewise be planted in warm regions. Oil from myrtle and laurel seeds should also be extracted now.

Further, January's waxing moon is the preferred time for planting rosebushes. In all grafting processes, but especially at this time, covering the graft with cow manure combined with a bit of straw is highly recommended. Additionally, cherry tree shoots should be planted at this time; cherry tree pits, however, are best sown in early December. Since cherry trees thrive in cold regions, they should be planted during cold weather, particularly if started from pits. Late-season green barley should be sown now, too. All work performed to generate growth, such as planting and sowing (other than the exceptions noted earlier), should be completed during the waxing moon because warm weather helps them take root and thrive. This is also the appropriate time to construct fences, which should be done by wrapping brambles and other plants in ropes.

JANUARY'S WANING MOON

According to Pliny, this is the time to gather remaining seeds and fruits, shear sheep, castrate livestock, prune vegetation, clear the ground, and store the seeds and fruits that have been gathered. (A general rule applicable to the waning moon during *all* the months of the year is that everything lasts longer when stored during the waning rather than the waxing moon.) In warm, early-harvest regions, vineyards should be pruned during the waning moon, provided they are protected from freezes. Pruning should be performed on clear, calm days when the vines have thawed. The process should be started two to three hours after dawn and finished before the next nighttime freeze sets in. All trees not previously pruned should likewise be pruned in accordance with this method, so that all trees will have been cut before they blossom, or before their buds swell.

During the waning moon, as well as during the waning of the day, wood should be chopped for building purposes, because then it will be well cured. Also, stakes for vine supports should be prepared, since they will then be durable.

Regarding crops, garlic and onions should be sown at this time to make them

less potent than they would otherwise be. Scallions, trimmed and with some of their smaller roots and a bit of the onion retained, should be soaked for a day and then planted in moist, irrigated soil. Planting scallions along with their tender roots produces sweeter fruit than if sown from seeds. Onion stems can also be planted using this method.

Trees, vineyards, and gardens should all be fertilized with well-ripened manure. Urine combined with ashes should then be poured into trenches dug at the base of trees and grapevines. Fertilizing trees at this time is critical, especially in cold and humid regions, whereas in hot, dry areas, fertilization should have been completed earlier. The trenches dug around grapevines in cold regions allows the roots to collect water. The young roots on the surface should be dried, however, because then they will be more easily removed. Weeds should be eradicated from grain crops and all other cultivated plants to prevent the vegetation from perishing. An initial plowing at this time effectively removes brambles, weeds, and undergrowth.

Building stone walls during this month is likewise recommended, provided the soil is not frozen. If the soil is frozen, or overly compact, wait until February.

FEBRUARY'S WAXING MOON

According to the predictions of agriculturalists, if February is variable and hot—meaning first cold, then dry, and later rainy—it will be a year of great abundance. If the winter is moderately humid with substantial freezes, the summer will be mild and very humid. If the summer is hot and dry, the fall will be mild and humid.

There is no better time to perform agricultural tasks than during February's waxing moon, and it is a particularly good time to sow seeds. During the waxing period, quarterly grain crops and lentils should be sown in dry, temperate lands; seeds sown in moist soil at this time often rot. Hemp, flax, and peas should also be sown, for if they are sown in March they may be damaged. Both varieties of barley should be sown now, as well.

It is also a good time to plant vine and tree shoots of any variety, for they have not yet sprouted, and to prune grapevines. Plant grapevines using any method, and graft them in early-harvest regions or temperate lands, or in irrigated areas. Plant osier, olive, willow, poplar, and other similar tree cuttings. Plant the aforementioned pitted fruit trees too, especially in cold regions. Transplant laurels and late-blooming pear and apple trees, as well; the early-blooming varieties should have been transplanted in January. In warm areas, these trees should have been transplanted before winter set in.

During February's waxing moon, oats and legumes should be sown, particularly in cold regions. Plant cane fields, preferably in humid areas, after soaking the cane roots for a day. Sow citron, lime, and orange tree seeds; also sow sweet basil, anise, dill, cabbage, mugwort, fennel, watercress, lettuce, leeks, *yerba santa* (mountain balm), parsley, mustard, poppy, summer squash, and other vegetables.

Good fences can be constructed at this time as well, using plants covered with old ropes. It is also the appropriate time to buy sheep and other livestock, because they are no longer endangered by the winter's cold, unless March is harsh. In fact, geese, chickens, and peacocks should now be incubating their eggs.

This is the appropriate time for all *de pasado* grafting. It is also the proper time to graft pear, apple, and other trees that generally bud during February, using the *de coronilla, de barreno, de hendido* ("cleft graft"), or *de junta* technique. Grafting should be performed before the tree tips begin to swell. Grafting any tree variety using the *de pasado* method is also recommended at this time. During the waxing moon, fragrances and medicines should be grafted into trees and vines before they begin to bud, as described previously. Verify that the tips are beginning to become plump before initiating the grafting process. Plant myrtle, pomegranate, and other late-blooming tree cuttings in early-harvest regions, while in late-harvest regions and cold lands, tree cuttings are best planted in March, or preferably April. During February's waxing moon, also transplant rosebushes in early-harvest regions.

Plant violets, lilies, saffron, and asparagus at this time, as well, although the latter

can often be successfully sown in January. Asparagus seeds, which thrive in humid climates, should be sown in plowed soil, no deeper than three inches. Later, their roots can be gathered and planted somewhat deeper, preferably in fallow ground. Ashes invigorate the seed, fostering healthy sprouts. In early-harvest regions, mulberry bush cuttings and sprouts can successfully be planted during this time, too. Since they are late-blossoming trees, however, it is preferable to wait for March's waxing moon, unless they are being planted in very hot regions in an early-harvest year. Additionally, shelled hazelnuts should be sown during this time.

It is a good time to buy pigs too, especially in early-harvest regions and hot areas where there is an abundance of weeds. If weaned pigs are allowed to graze now, they can improve the land considerably.

There is commonly more of a northwest wind during this month than at any other time, and it infuses the land with substance and temperance. If temperatures are not excessively cold, all types of trees and flowers could be planted before they sprout leaves. The roots of mint, other herbs, and vegetables should be planted as well, to generate new sprouts. Plant *uvillas del paraíso* (small paradise grapes) and irrigate them thoroughly. This is also a favorable time for planting cuttings of other varieties of paradise grapes, whose leaves resemble olive leaves in shape and color, except that they are whiter. Both jasmine and carnation shoots can be planted to create flower beds. Their seeds can also be sown, as can those of poplars, cypresses, *peros de nelda* (nelda pears), and similar fruits. Also consider transplanting poplar, cypress, willow, ash, and olive trees. Finally, all tree seeds, such as ivy, myrtle, and laurel, should be sown during this period.

FEBRUARY'S WANING MOON

During February's waning moon, in the morning when it is cold place well-ripened manure and rotten urine in the trenches surrounding late-blossoming trees and grapevines so they will produce more numerous and tastier fruit. They, like all trees, should be fertilized before winter as well. Prune vineyards in temperate lands, and tie and prop grapevines on trusses before the buds begin to swell. If this is done at another time, grapevines may perish or be damaged, produce less fruit, and be poorly trussed.

In addition, this is an excellent time to plow fields for sowing. Plowing and digging trenches around grapevines is highly recommended for killing weeds, particularly in areas where the soil is rich. If the weeds grow back, clear the area again to eradicate them—ideally in May in temperate or cold lands, and in April in warmer regions. When eradicating weeds from vineyards this late, plows should not be used, because they may destroy many of the young, delicate tendrils.

During February's waning moon, as well as January's waning moon, cane is ripe for cutting, making the wicker available for weaving baskets and other items. Wicker for white weavings, however, should not be cut until it is fully grown, since the moisture it produces while maturing facilitates stripping the bark.

This time is also propitious, especially in cold regions, for removing young roots and sterile shoots from grapevines and small trees, if they were not taken off during January's waning moon; in warmer areas, this task should be done earlier. Late-blossoming trees should be pruned so their potency is not sapped by the branches. Likewise, remove dry and excessively long branches, and eliminate all worms and waste.

Throughout this month, pigeons are beginning to bear their young, hence delousing should begin in order to control lice infestation. If the weather is pleasant, and preferably before hatching begins, honeycombs should be removed from beehives, as well, eliminating any dry or rotten sections. Beehives should also be protected, because hibernating bears, awakening quite hungry at this time, can cause significant damage to beehives. Because bears quickly flee when they smell sulfur, place rags soaked in sulfur near beehives in bear habitats, making sure they are hung so the wind carries the sulfuric odor throughout the area. This is also a good time to infuse the beehives with rosemary and other fragrances, to attract and nourish the bees if flowers are unavailable—a task to be continued throughout the winter in cold and sterile regions.

Beginning in February, although March is preferable, cages and traps should be prepared for young rabbits and wild boar. During February's waning moon, fences can be constructed in cold and humid areas, and livestock castrated in early-harvest regions.

MARCH'S WAXING MOON

Certain tasks can be performed either this month or in February or April, since every month falls between two others having similar qualities. This is particularly true of February and March, because they have some of the cold and precipitation of winter, combined with the heat of summer. Although performing tasks at the indicated times is preferable, doing them fifteen days earlier or later is quite appropriate, except for tasks related to grafting, which are more successful when undertaken within the prescribed time limits.

In late-harvest regions and frigid areas, this is the proper time to plant and tumble late-blooming grapevines, while early-blooming grapevines are best planted in January and February. Plant grapevine shoots, preferably those removed after the grapevine sprouted, since they take root better. This is a favorable time for grafting vines of other sorts as well, and for planting myrtle and mulberry tree cuttings. In irrigated lands, it is also beneficial to plant elm tree cuttings that have been removed during the latter part of the day.

Fields with rich soils should be guarded to foster the growth of thick grass. In warmer regions, this is a good time to sow millet and broomcorn millet, also known as panic grass, especially if the land is irrigated. Similarly, it is the appropriate time to sow pomegranates, hemp, and flax.

Further, grapevine stocks can be split to generate new branches, and damaged grapevines can be treated. In warm regions, these tasks should be performed in February, well before budding season. This is also an appropriate time to sow alfalfa,

peas, melons, snake cucumbers, ordinary cucumbers, squash, and thistles. In addition, it is a good time to gather soil around trees in warm and early-harvest regions, and to sow asparagus in cold areas. And rue, coriander, anise, dill, chard, wild amaranth, and other similar seeds, as well as those of sour fruit trees, are sown now for summer. It is the proper time to transplant pomegranate cuttings and to sweeten seeds in milk or honey water for sowing. Abencenif claims that fruit from citron trees, orange trees, and other sour fruit trees will be less sour if they are hoed and a hole made in the trunk is filled with sugar. Figs should now be planted in temperate regions, while in warmer areas they should have already been planted. During this period, late-blooming pear and apple trees can be successfully grafted, as well. Wild pears, called *servas* in Spanish, can also be sown. And it is a good time to plant *azufiafo* (jujube) cuttings or trees.

Additionally, this is an excellent time to purchase cattle, because they are not yet corpulent and, due to their smaller size, are more easily tamed and broken. In addition, new traps and cages for rabbits and other animals should be constructed, or old ones fixed. During this month, bees labor among the flowers and fresh elm tree leaves, though plant milk that has been collecting can do them much harm. Assist the bees according to the suggestions provided earlier, and spray and cleanse the area. Also allow for the construction of new beehives by the swarms that vacate their old hives. Throughout this month, visit beehives often, because the swarms are just developing and it is important not to lose the initial swarms for they are superior to the others.

MARCH'S WANING MOON

During this waning moon, it is advisable to prune vineyards in late-harvest regions. They are well seasoned for pruning at this time, since they bleed far less, do not freeze, and their buds do not burn. Pruning should never be performed later than this waning moon, because vines are already budding and branching considerably, and if pruned later, will have less strength and more likelihood of burning. People who neglect their vineyards in March validate the saying *"La viña del ruin se poda en abril!"* ("The bad vineyard is pruned in April!")

Now is also the appropriate time to hoe new grapevines in late-harvest regions, while in early-harvest regions this task should be done in February. Vineyards should be excavated now as well, in order to prevent harm to the buds that will soon form, and new grapevines should be trussed, tied, and propped. This is also the indicated time for plowing grain fields in preparation for sowing and to prevent weed growth. Flax and hemp fields are plowed a second time in cold areas, preparing them for sowing during the next waxing moon. The fields should then be irrigated. Trees and gardens also should be irrigated in late March.

In colder regions, this is the recommended time for pouring unsalted olive juice and water into the trenches dug around olive trees and for making holes in sterile olive trees to insert very stiff olive wedges. Similarly, it is a good time to trim the tops of mulberry bushes and fig, pomegranate, and other late-budding trees. Effectively protect gardens, as well, so livestock do not graze there or trample the stalks that are just beginning to form at this time. Also when temperatures are cold, wine should be decanted and stored by either burying it or placing it in a wine vault, if such tasks have not yet been done.

Further, this is a good time to perform tasks related to sweetening bitter almonds. Almond trees can be irrigated with human urine now so they produce sweeter almonds. Another method for sweetening bitter almonds is to shell them in hot water according to the instructions for producing almond milk, then place them in a cask filled with sweet water, allowing them to soak for nine or ten days, changing the water daily. Almonds sweeten more rapidly if placed in running water; in fact, a basket of bitter almonds placed in a rapidly flowing river will, within three days, become as sweet as any other variety. And while almond milk, nougat, and other products can be derived from almonds sweetened in this way, the oil is unusable.

APRIL'S WAXING MOON

Mulberry, pomegranate, and boxwood tree cuttings can be successfully planted this month in cold, irrigated regions, while in warmer, unirrigated regions, they should be planted earlier. Fresh olive tree cuttings too can be successfully planted in cool, irrigated lands. Alfalfa should be sown now in cold regions, as well. Although most garden plants can be sown any time between January and August in areas where they grow successfully, April's waxing moon is a good time in most regions to sow melons, cucumbers, leeks, capers, scallions, coriander, squash, peppermint, celery, and lettuce. It is a favorable time in warm areas to plant jujube tree cuttings as well, and those of other young trees, while in colder regions they can be planted as late as May. This period is equally propitious for grafting olive trees using the *de escudete* or *cañuto* technique. Peaches and *priscos,* too, can also be grafted using these methods. If the region is hot, citron, orange, and other trees are generally successfully grafted within their species as well. The *de escudete* method can now be used to graft fig trees in hot or dry regions, but such a task is best performed in May and June. This is also a favorable time for grafting mulberry bushes, jujube trees, and service trees, although the latter can be grafted as late as May.

Calves graze during April, and if sufficient grass is unavailable their mothers provide adequate nutrition. Also during this month, continue to ensure that bee swarms do not relocate and protect bees from butterflies by lighting afternoon fires between the beehives. Sheep and goats impregnated during this month's waxing moon give birth early, resulting in lambs and kids that are large and strong when the winter cold arrives. In cool regions where hens incubate their eggs in April, the chicks hatch by May, as opposed to the summer months, when newborn chicks are often afflicted with disease-induced blindness, causing their heads to swell and resulting in death in the winter.

APRIL'S WANING MOON

This is the recommended time for plowing rich and moist soils in warm regions, where abundant rains have recently softened the soil. Plowing allows the sun to penetrate and dry the soil's harmful moisture. Vineyards, however, should be dug and trenched with caution, because buds are quickly sprouting. In still warmer regions, trenches dug around trees and grapevines should be covered now, while in hot and dry areas grain fields should be irrigated.

Also during April's waning moon, sheep should be sheared in warm regions. In addition, beehives should be cleaned again, ridding them of the numerous small insects and spiders that accumulate on them. Because beehives should always be well covered with clay, it is essential for their occupants to continually eliminate any holes that may form. "Bees' work is never done," as they say, and the holes simply create more work for them.

MAY'S WAXING MOON

During this month, in cold and late-harvest regions millet, broomcorn millet, and peas are commonly sown. In warmer and early-harvest regions, peach, *prisco*, apricot, almond, orange, citron, lemon, early-bearing fig trees, and olive trees can be grafted using the *de escudete* technique. Cabbage and melons also can be sown at this time. While leeks can be successfully transplanted now along irrigated furrows in warmer regions, this is best done in the fall. Purslane too can be sown now, although it sprouts only in hot weather. Pomegranates to be grown in a large cask are best sown at this time, as well.

In early-harvest regions, it is the time to cover the roots of the stocks and trees that have been excavated, if this task has not yet been done. It is also the best time of year to sow sour fruit seeds, especially in cooler regions, where they sprout more quickly, grow stronger, and are in less danger of freezing. Those remaining in their fruit until they are sown maintain their tenderness and sprout more quickly.

During May's waxing moon, butterflies that converge on beehives must be exterminated, as before. In hotter regions, male goats and rams should be placed together with the females to impregnate them so they will give birth at a favorable time of year.

MAY'S WANING MOON

Leaves should now be removed from grapevines [according to the specifications set forth in chapter 2], eliminating whatever might damage the stocks or hinder the development of fruit. In cold regions, it is also time to plow through rich and moist soil, preferably when the sun is shining, to destroy the weeds. Plowing is likewise recommended for gardens sown in the fall and for lupines sown for purposes of fertilization. During this month, olive trees should be pruned in late-harvest regions, although outcomes are usually more favorable if this task is performed earlier. Also, irrigated trees should be watered, although not in cold, late-harvest regions.

In early-harvest lands, this is a good time to harvest hay before it dries. Also in early-harvest lands, even barley should be harvested if it appears ready, since when it becomes too dry, much of the grain is either lost or distinctly less flavorful. Beginning in the latter part of April and lasting throughout May, vineyards should be frequently examined for aphids, which tend to proliferate in places with excessive moisture. May is also a good time to castrate steers, piglets, and lambs, and shear sheep, particularly in colder regions.

June's Waxing Moon

Millet and panic grass can be sown as late as June in cold regions. All trees with thick, flexible bark, such as fig, olive, orange, citron, laurel, and almond, can be effectively grafted using the *de escudete* method. This technique can also be used to successfully graft plum trees into almond, peach, or other similar trees. Moreover, borage, cabbage, and additional garden vegetables can be sown during this time in well-irrigated early-harvest areas. This is also the appropriate time to slightly curve the branches of pomegranate trees to stimulate their growth.

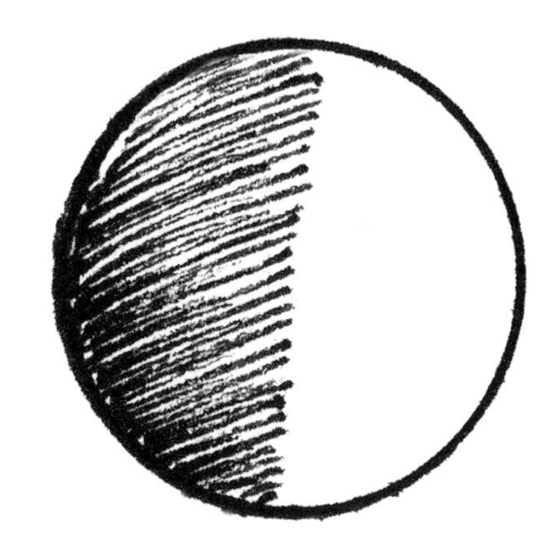

June's Waning Moon

During this waning moon, cease irrigating fig trees that are already bearing fruit and the fruit will ripen more quickly and be less watery. Also, the threshing floor should be prepared for harvest time [as explained in chapter 1]. Beans, chickpeas, and other dry legumes can be harvested. It is time to reap barley as well, and in hot and early-harvest lands, wheat. In addition, reap hay pastures, and in cooler regions remove flax and hemp. Also in cooler regions, if rain has softened the vineyard soil it should be plowed to eradicate weeds. Grain fields too can be plowed, particularly in cold and late-harvest regions. If the fruit has grown densely on both early-ripening and later-ripening trees such as pear and apple, it should be thinned so the remaining fruit will grow better and more abundantly and the trees will not age as quickly. Removal of the fruit should be performed earlier, if necessary, and is more beneficial where trees are not generally irrigated.

Further, this is the recommended time to shear sheep in cold areas, since the wool is enhanced by the sheep's perspiration. New swarms of bees are still forming now,

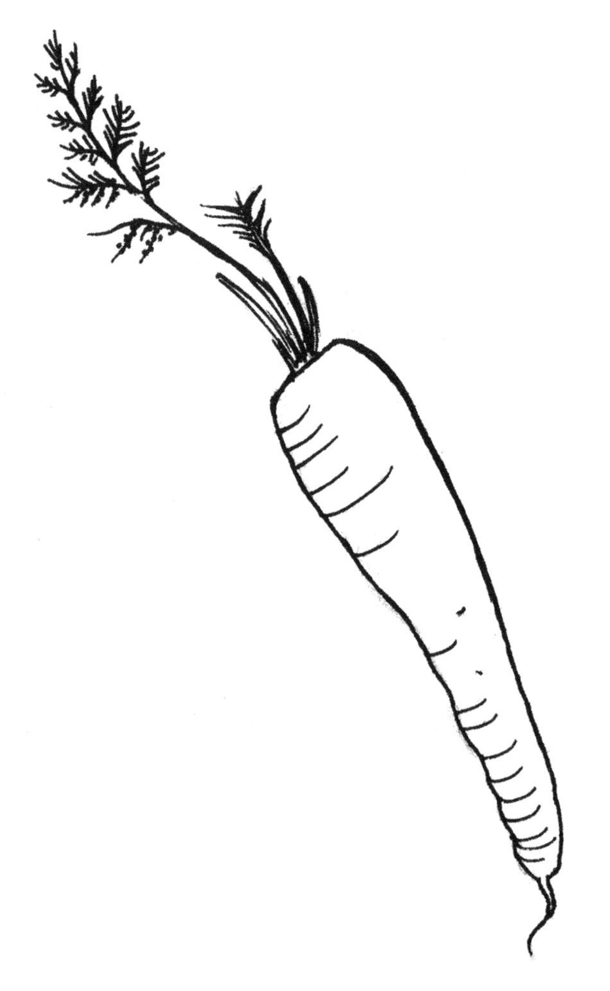

especially in late-harvest regions; and everywhere, beehives should be sterilized to improve the quality of the honey.

JULY'S WAXING AND WANING MOONS

The most important task during July is to complete grain reaping and gathering in late-harvest regions, preferably during the waning moon, ensuring that the harvested grain will be better preserved. If vineyards are located in warm, dry areas, the stocks' roots should be well covered to prevent the sun from desiccating them. Turnips, carrots, cabbage, onions, and other winter vegetables should now be sown, along with mustard. If the region is cold, trees can still be successfully grafted using the *de escudete* method. This is also the time to remove weeds and ferns, because those cut now will not grow back.

During this time, it is beneficial to have dust in the vineyard, which can be effectively produced by excavating the soil at the base of the stocks in the mornings and again in the afternoons. The loose soil stimulates grape growth, accelerating the ripening process particularly in humid regions; in warm, dry areas, the soil can be used to cover the stocks' roots, preventing the sun from desiccating them. Where grapevines are not irrigated, soil that cracks open due to dryness should be well excavated around the roots, preventing the sun from penetrating the roots' fissures and causing harm, particularly to young vine shoots.

Citron and lemon tree cuttings can be transplanted as late as this month if they are well irrigated, although it is preferable to complete the task earlier. If the bases of these and other fruit trees have been excavated in regions where the grain has been reaped, they should now be covered to protect the roots that were once shielded by grain stalks. Likewise, fruit that ripens in the fall should be removed from trees that are carrying too heavily. Almonds, at the very least, are now ripe for gathering.

During July, or in some places June, bulls should be allowed to impregnate cows so they give birth at the proper time. Also during this month, female sheep and goats should be placed together with the males for impregnation, since lambs and kids born early are stronger and healthier, with reduced risk of dying. Finally, pigs and other livestock should be allowed to consume the *rastrojo* (stubble) and fallen bits of grain in the already reaped late-harvest hay pastures.

AUGUST'S WAXING AND WANING MOONS

August's tasks are remarkably similar to those of July. In addition, this is good a time to dig wells, because if it fails to rain, the intense sun dries everything, and underground water tapped at this time then becomes a guaranteed source of water for the entire year. This is also the recommended time to burn pastures and fertilize grain fields, replowing them and covering the fertilizer in preparation for sowing. Lupines are sown at this time, as are late-ripening turnips, radishes, and cabbage, especially if it has rained. Alfalfa and alfalfa seeds should be gathered at the end of this month so they will be ripe for sowing later. In humid regions, grape leaves should be removed from the vine, allowing the bunches to receive sunlight, and the branches should be lifted so the grapes do not spoil. In dry regions, the bunches should be covered to prevent desiccation. In all areas, but especially early-harvest regions, all necessary preparations must now be made for the grape harvest.

During this month, weeds and ferns can be eradicated by plowing and removing them. Further, figs, peaches, *priscos,* and plums can be dehydrated. Finally, at this time large flying insects fervently pursue bees and, if possible, should be exterminated to protect the production of honey.

SEPTEMBER'S WAXING MOON

During this month's waxing moon, sowing should be initiated, particularly in cold areas with thin soils, where seeds such as rye and barley, harmed by an abundance of precipitation, are in danger of failing to sprout properly. Peas, beans, and lupines are sown during this time in warmer regions, because they fail to sprout adequately in cold weather. In addition, pasturelands are plowed, cleaned, and fertilized with fresh manure. And in rich soils that have been properly worked this is the appropriate time for sowing wheat, especially white wheat, because of the advantages described [in chapter 1]. The commonly irrigated variety of flax called *bayal* can be successfully sown at this time, as can green barley and meslin. Garden poppies can usually be sown in warm regions. Finally, pink carnation shoots can be successfully planted at the end of the month.

SEPTEMBER'S WANING MOON

Grape harvest preparations in humid and late-harvest regions must be completed at the beginning of this month's waning moon, if not sooner. First, this requires tagging grapevines that do not produce fruit or ripen properly, as well as fruitful plants, identifying vines that need to be grafted and those that can be used for grafting and transplanting. In these humid regions, where grapes commonly spoil, it is best to remove leaves from the grapevines, particularly those surrounding grape bunches, allowing the grapes to dry; they should then be harvested in case of rain. Next, gather the vine shoots with unripened grape

bunches and hang them on trusses for storage in a sheltered area such as a portal, thus protecting them from cold and rain, both of which are damaging to grapes. This is also the proper time to gather and store alfalfa for the winter.

Further, this is the recommended time to fertilize lands to be sown, preparing them thoroughly to absorb the first rain of the planting season. In hot and dry regions, it is also the appropriate time to carry out preparations for planting trees before winter, digging holes and selecting and tagging trees for cuttings. And plowed lands to be sown in the spring with garden vegetables or three-month seeds should now be excavated.

Beehives that were not sterilized in June should be sterilized now, particularly in regions that are hot and temperate. If the beehives are overflowing with honeycombs, remove a portion, leaving the rest for winter nourishment. Beehives in hot and temperate climates can be sterilized in October.

OCTOBER'S WAXING AND WANING MOONS

Grapevines should be tagged at the beginning of this month in dry and late-harvest regions, then the ripe grapes should be harvested and the vines excavated in such a way that their leaves fall into the trenches surrounding their bases so the vines can better absorb water throughout the winter. In hot or temperate lands, prune thin grapevines after they have lost their leaves, and be sure to remove their spikes.

At month's end, dig trenches around the bases of trees (*atetillars*) and plant poplar, willow, palm, and olive tree cuttings. Also transplant almond and all early-bearing fruit trees. In warm regions, if it is raining plant cherry, early-blooming pear, and apple trees, along with trees that thrive in the cold. In cooler regions, use dirt to cover the trenches that surround the bases of citron, orange, and similar trees, as well as the trees themselves, protecting them from freezes. While grapevines and trees can be grafted at this time, they are not usually successful. For better outcomes, holes should be dug to plant trees and grapevines in the spring, placing manure and water inside them to decompose and fertilize the soil.

Further, this is the ideal time for gathering produce, sowing seeds, and planting crops other than fruits and nuts. Gather acorns, chestnuts, walnuts, and hazelnuts; also gather quince and other late-blooming fruits. From mature olive trees, extract green olive, an excellent ingredient for food preparation, especially sautéing, stir fry cooking, and salad dressings. Sow grain, such as wheat, rye, barley, and flax, as well as broad beans, peas, lupines, and other legumes. If the soil has not yet been plowed and fertilized, these too are tasks best completed now, particularly for planting seeds such as mustard, dill, chives, capers, spinach, and other vegetables that thrive in sunny areas. If it rains, thistles should be covered with soil or transplanted. Other crops to transplant now include leeks in their furrows, peppermint roots, and irises.

NOVEMBER'S WAXING AND WANING MOONS

If the sowing for spring has not been completed, wait no longer. During this month, in dry, hot climates plant trees that thrive in cold weather; other trees are best planted in January or later. In humid lands, plant cane fields. In warm areas, plant vineyards, tumble the heads of vines, remove shoots, and plow the soil to eradicate weeds. Allow pigs to roam through the vineyards, grazing

on the grass and all the while excavating the land. Vineyards, and trees as well, should be fertilized and cleared of dry branches. Trenches should be dug around olive and other large trees that do not commonly freeze, so they can properly absorb water. This is also the recommended time to plant garlic, particularly white garlic. Now and during December's waning moon, excellent jerky can be produced, and wood, cane, and wicker can be prepared for building.

DECEMBER'S WAXING AND WANING MOONS

Throughout this month, working in fields should be kept to a minimum, since the soil is hard and can be severely damaged. Although most tasks will be performed in and around the house, fences can be repaired in the fields, ditches cleaned, fields fertilized as necessary, and supports for vines constructed. If the planting for spring has been postponed, that too can now be accomplished, including the sowing of garden vegetables such as lettuce, radishes, and garlic. In addition, trees can be fertilized and urine poured into the trenches encircling them. It is also possible to plant walnuts, chestnuts, and acorns now, although this is best done in November or January.

During this month's waning moon in particular, it is advisable to chop wood and collect manure in new dung heaps. It is likewise a good time to fashion tools, season barrels, clean casks and wine vaults, and hunt.

Weather Indicators

God provided us with countless celestial and terrestrial indicators of weather conditions to aid us in our agricultural endeavors. While many of these indicators are quite apparent, they may be difficult to recognize due to our limited understanding of such phenomena. Yet in failing to recognize them, we falter needlessly.

The majority of indicators discussed in this section are observable signs that everyone who works the land ought to be able to identify and use beneficially. It should be noted, however, that signs in one region may not hold true for other regions. So that farmers can better predict weather conditions based on well-recognized indicators, I have gathered the knowledge that follows from classical sources: Aristotle, Pliny, Virgil, Ptolemy, and an anonymous treatise entitled *Mutatione Aeris*, Latin for *Weather Changes*.

RAIN, STORMS, HAIL, FREEZES, AND SNOW

If the sun appears fractured when rising, this is an indicator of rain.

If the sun rises among red, black, or gray clouds, it is a sign of rain.

A halo surrounding either the sun or the moon indicates wind and rain.

Scattered rays of sunlight or clouds toward the north or south as the sun rises indicates rain and wind are imminent, despite a seemingly calm sky.

If the sun's rays are obstructed when it either rises or sets, it is a rain indicator.

Lines of rain in front of the rising sun, even if no clouds are present, is an indicator of impending rain.

The rising sun emitting extended rays between clouds, even if the day is somewhat clear, indicates rain.

If the sun emits some rays before it rises, it signifies rain and wind.

If there appears to be a dual sun and clouds become reddish, particularly in the morning, it predicts rain.

Dark rings surrounding the moon indicates rain.

The rising moon encircled by a dark halo indicates rain.

If the new moon has a higher and darker horn above it than below it, this signifies rain during the waning moon.

If the lower horn of the moon is higher and darker than the upper one, it predicts rain during the waxing moon. If the moon is dark in the middle, it indicates rain during the full moon.

A new moon in the south forecasts rain.

The weather during the quarter moon remains throughout the cycle, regardless of whether it is rainy, windy, or calm.

When encircled by halos, planets—which are easily discernable among stars because they do not twinkle—are a sign of rain. Stars encircled by halos also indicate rain.

A double rainbow is sign of rain, even if it appears after a downpour.

A noonday rainbow signifies heavy rain. A rainbow appearing in the west suggests thunder and light rain are imminent, while appearing in the east it indicates calm weather. A cold wind combined with a rainbow also forecasts calm weather. A double rainbow signifies rain, whereas a double rainbow appearing after a downpour signifies impending calm weather.

When the sky is calm and lightning suddenly strikes, there will be rain, thunder, and cold weather.

More lightning than thunder in summertime is a rain indicator.

Thunder at noon predicts rain, and thunder during the winter signifies continuing rain and very high humidity.

White clouds at sunset, resembling elongated masses of wool, indicates rain within a few days.

Clouds rising above mountain peaks ensures imminent rain.

When lit candlewicks begin to flicker, or embers stick to pots, it signifies impending humid weather.

Chimney soot falling swiftly and abundantly is a sign of continuous humidity and rainy weather.

Moist ashes collecting in a fire are a sign of rain.

Frogs singing more than usual, and pigs frolicking and creating havoc, forecasts rain.

Worms and caterpillars emerging from underground, and ants hastily protecting their food or their eggs, indicates rain.

Small bubbles on rainwater are a sign of additional rain.

Birds removing their fleas are a sign of rain.

Swallows flying so close to water that they almost touch it with their wings predicts rain. Crows cawing loudly, then stopping and beating their wings, means rain is imminent.

Bells or any metals clanging more clearly than ordinary is a sign of rain.

Clouds moving toward the east or north from the south or west signifies continuous rain.

Wind blowing cold and hard after a downpour, is an indicator of recurrent rain. People with head injuries, limb injuries, or gout can effectively predict rain and humidity when exposed to this indicator.

Red clouds in both the west and east is a sign of rain.

The greater the number of clouds obscuring the sun, the heavier the downpour.

Dark, harsh, ominous clouds during sunset forecasts a storm the next day.

Many clouds forming before sunrise indicates a storm the next day.

A white halo surrounding the setting sun signifies a storm, and if fog is present it will be a great storm.

The rising moon encircled by two halos indicates a storm, as does a new moon with fat horns or an apparently swollen moon in mid-month.

If the sun appears to be enveloped by a white halo shortly before rising, there will be a small storm. Combined with fog, this forecasts a major storm.

Stars losing a bit of their clarity indicates a powerful storm.

Lightning throughout the sky ensures an impending storm.

Geese honking excessively are sometimes a sign of a storm.

A yellow fire or candle flame that crackles indicates a storm.

Food plates and bowls that leave moisture on a tablecloth are storm indicators.

When the sky becomes ominously akin to fire, it forecasts a storm.

A bright red sun emitting burning heat is a sign of either hail or rain.

Afternoon thunder and lightning guarantees a downpour or a hailstorm. The later in the day the thunder and lightning occur, the worse the storm will be.

A yellow rising sun during the summer suggests it will hail that day.

Hail in the fall or spring guarantees a freeze, endangering all trees, particularly fruit trees, unless a substantial amount of sunlight follows the hail, drying and purifying the land and air.

Hail in June indicates considerable frigidity in the higher atmosphere, and if it does not rain soon thereafter, the winter will be very cold.

Cold, dry weather without freezes suggests it will soon snow.

Herons sitting on sandbars far from the water, and seemingly sad, indicates an impending weather disturbance.

WIND

Red clouds that appear before sunrise indicates an impending wind.

If the rising sun has a halo, it is a sign of wind, which emanates from the direction of the part of the halo that first appears. If the halo then disappears, the day will be calm.

Often winds at high altitudes blow in the direction opposite the wind at lower altitudes, but occasionally the wind's direction above and below is the same.

An intense purple halo when the sun rises or sets suggests high winds coming from the direction where the halo first appears.

A red-colored rising moon is a sign of winds.

A new moon with enlarged horns, or with one horn pointing toward the Borealis or north, forecasts wind.

A moon with a fading halo indicates winds emanating from the direction of the initial fading, or where it is clearer.

If the setting sun is enveloped by a dark halo or fog, it indicates wind.

Flashes of light soaring across the sky, akin to falling stars, predicts wind—a few flashes indicating slight wind, many flashes signifying intense wind. If these flashes emanate from numerous directions, the wind direction will be changeable.

Morning thunder indicates wind.

When the amount of thunder exceeds that of lightning during the summer, wind will emanate from the direction of the thunder.

Fire or candle flames flickering erratically indicates imminent wind, while if there is no flickering there will be no wind; a flickering flame suggests a variable wind; and a flame moving in one direction indicates the wind will originate from the opposite direction.

A small cloud suddenly appearing when the sky is calm, particularly if it comes from the south, predicts a mixture of wind and rain.

CALM WEATHER

A clear sun that seems to sparkle less than usual signifies a calm day, particularly if the previous sunset was calm and clear.

Clouds turning red in the west at sunset indicates the next day will be calm. As it is commonly said, "*Esta noche arreboles, mañana habrá soles.*" ("Red sky tonight, sunny tomorrow.")

Clouds moving west at sunrise predicts calm.

A clear rising moon indicates calm weather.

Clouds surrounding mountains and migrating toward the valleys, leaving the peaks exposed, ensures calm weather.

Seasonal Predictions

In general, the predominant characteristics of one season cause the opposite to occur the following season. For example, dryness during one season causes humidity and precipitation to prevail during the following season, and vice versa. Similarly, ordinarily humid winters with heavy precipitation are followed by calm, dry summers; and calm, dry winters signify humid summers. Calm falls forecast windy winters. Heavy precipitation in spring and summer indicates a calm fall, but if the previous seasons are dry, the fall will be quite humid.

With regard to predicting the size of a fruit harvest, winters with extreme freezes, particularly in January and February, forecast an abundance of fruit and wine, since the trees and grapevines have yet to bud. Winters that are mild and gentle, on the other hand, can cause fruit trees and grapevines to bud early and consequently freeze. And with respect to using seasons as a guide to agricultural occurrences, it is better not to divide the year into seasons as astrologers generally do, but to focus instead on cycles of the moon.

Also more important than abiding by the traditional division of seasons as an indication of weather conditions is the willingness to pay attention to natural phenomena that influence the weather regardless of the season. For example, when migrating birds such as cranes, pigeons, and thrushes arrive early, it is an indication of an early winter, and when these birds migrate late, it signifies a delayed winter. When they return late, it indicates a cool summer. If they leave early, it forecasts a hot summer. Birds such as turtledoves, quail, and sparrow hawks, migrating early from warm climates to summer in this area indicate an early, hot summer. A delayed arrival indicates a late, cool summer.

An additional bit of useful information is that the first half of a month will generally have weather similar to the last half of the proceeding month. As it is commonly said, "*Cuando un mes desmedia a otro semeja.*" ("Conditions during the latter half

of the previous month, coincide with those of the first half of the following month.")

Finally, because weather changes vary according to the nature and conditions of the land and indicate somewhat different possibilities in different areas, it is vital to have an understanding of the common causes of weather patterns in particular regions. This requires knowledge of the consequences of various winds, the interactions of certain types of clouds with local landscapes, and other meteorological factors contributing to the weather patterns forecast by local indicators. Even so, the preceding signs can be used as a general guide to weather prediction in most areas.

May God grant us health so that with it we may serve him. Amen.

GLOSSARY OF TRADITIONAL AGRICULTURAL TERMS

Primary sources for the information that follows include *Tesoro de La Lengua Castellana o Española*, compiled by Sebastián De Covarrubias Orozco in 1611, revised by Manuel Camarero (Madrid, Spain: Editorial Castalia, 1995) and *Diccionario de la Lengua Española* (Madrid, Spain: Real Academia Española, 1992).

ABREVADERO A watering hole, from the French words *brevis,* meaning sheep, and *abreviar,* to hasten, used to describe camels approaching an oasis. Eighteenth-century documents regulating water use in New Mexico stipulate that an acequia cannot be used as an *abrevadero.* Even along rivers, livestock were allowed to drink water only at specified places where they would not pollute the downstream flow.

ACEQUIA MADRE A primary channel delivering water and thereby giving life to the land below it, from the Arabic *as-saqiya,* meaning that which gives water; more specifically, mother ditch. This type of waterway follows the contours of the land and, like a vein carrying blood to the human or animal heart, delivers nourishment to a high-desert landscape. Every *acequia madre* forms its own terrace on the land; where one ends, another begins, forming a separate terrace.

ACEQUIAS MENORES Secondary ditches, known also as *acequias secundarias,* that run perpendicular to an *acequia madre* and take water to more than one *parciante,* or piece of land. These ditches are under the jurisdiction of neither the *comisión* nor the *mayordomo* of the acequia madre. They are widely referred to as *linderos,* from the Latin *lindes,* meaning contiguous, and also described as *brazos,* Spanish for arms. A secondary ditch that runs horizontal to an *acequia madre* is called a *cabezera*.

ALBERCA A small pool that collects water for later use in irrigating gardens *(huertas)*. Such pools have all but disappeared in New Mexico, but are still widely used in Spain. The word has Hebrew roots, as in *bereca*, Hebrew for pool, though the first syllable, *al*, suggests Arabic influences as well.

ALFALFA A plant widely grown for hay or forage, from the Arabic *al-fasfasa*. This plant is known in Spanish as *yerba medica*, having originated in a region of Asia known as Media, which was named after Medo, son of the mythological Medea and Aegeo. See also *melga*.

ALMÁCIGA Known throughout the Spanish-speaking world as a seedbed, a place to start plants early. The term is thought to originate from either the Latinized Greek word *al-maciga* or the Arabic *al-maskaba*, an irrigated place.

ALMANAQUE A calendar containing astronomical and meteorological data for the year ahead. The word has many possible origins, such as the Arabic *al-manaj*, which derives from the Latin *manachus*, meaning months. Diego de Urrea, a source quoted by Covarrubias, traces the word to the Arabic *manaquebu*, from the verb *necabe*, meaning something that is forthcoming. Its roots are also associated with the Hebrew *Almanah perpetuo* of Rabí Abraham Zacuti, which comes from *manah*, meaning *numerare*, an astrology table based on numbers. The popular New Mexican *dicho* (saying) "*Eres como el almanaque Guadalupano, anuncias agua pa' hoy y cae mañana*" compares an unreliable person to the Guadalupano almanac, asserting, "You predict water for today and it doesn't come until tomorrow."

ALQUERÍA A small village, from the Arabic *al-qarya*. Also, the Spanish term *alcarria* refers to a series of villages or hamlets, each having a commons and land irrigated by an acequia. The *mercedes* (land grants) of New Mexico likewise call for a commons and land irrigated by acequias.

ALTITOS Highlands where fruit trees are planted, making them less susceptible to freezes than those growing by the river, where cold air tends to settle. Generally these tracts of land are immediately below an *acequia madre* and form the upper terrace of a *suerte*. Each one has a unique microclimate, which informed traditional landowners of the crops that it could or could not support.

ATOLE A blue-corn mush, from the Náhuatl *atolli*. Before the introduction of coffee, *atole* was the drink of choice among Mexican people. A well-known poem, called "*Trovo del café y el atole*," pits coffee as an outsider against *atole*, and in the end, corn conquers coffee.

CAÑADA A road used for moving livestock between winter and summer pastures, from the Latin *canna*, meaning wild road. This part of the commons, used traditionally to transport mostly sheep and goats back and forth from the lowlands to the mountainous regions, lies between two high peaks called *lomas* and *cuchillas*, has *abrevaderos* (watering holes), vegetation for grazing, and being at least ninety varas (eighty-two and a half yards) wide, *descansaderos* or *majaderos*, places where the animals can rest and also deposit manure. These roads were commonly called *cañadas reales,* or royal cattle paths.

CIÉNAGA Marshland or moor, from the Latin *caenica.* This land formation usually contains black mud that is smelly, soft, and, without drainage, able to be used only marginally for grazing. With drainage, it can also be used for growing crops.

CODO Elbow; also, the distance from elbow to knuckles, measuring approximately nineteen inches. Body-based measurements, common among illiterate farmers of pre-sixteenth-century Spain, also included *dedos* (fingers), measuring inches from the *dedo pulgar* (thumb), and *palmas* (hands), the equivalent of about four inches.

CHINAMPA A fence made of canes, from the Mexican *chinámitl*; also, a small island or floating garden for growing vegetables and flowers. Initially, these small garden tracts were moved from place to place, in keeping with the owner's wishes, but over time, as the water level dropped, they became stationary. Still in use today, the large rectangular beds of mud, constructed in the midst of various lakes, operate hydroponically: if too low, they become flooded, and if too high, the plants cannot absorb enough water. Probably the most famous *chinampas* are those in the Xochimilco area of Mexico City.

DEHESA A seminatural ecosystem requiring a certain amount of human involvement; also, an outdoor space that conserves a great number of flora and fauna. In Latin, the equivalent of a *dehesa* is a *pascua,* a place where livestock is grazed, which could very well have come from the Roman custom of establishing *latifundios*, or country estates, in marginal lands. According to Covarrubias, however, the term derives from the Arabic verb *dehesa,* which has to do with the thickening of the land due to the moistness and growth of weeds, making lowland soil that is difficult to walk on. But, he adds, it could also have evolved from the Hebrew *deshesa,* a parcel of land choked with weeds.

Dehesas today are considered agroforestry, or agrosilvopastoral, systems reserved for the joint production of trees and agricultural crops or animals. As before, they are known for their poor soil, harsh climate, and need for human intervention to make them somewhat productive. Those that once formed part of New Spain's land grants in Texas, California, and Louisiana became private property, while those in New Mexico remain under federal control and are now maintained by the Bureau of Land Management, the State Land Office, and the Forest Service. They are predominantly a type of pastureland with scattered evergreen trees such as piñon, juniper, and deciduous oak. The piñon trees, for one, are pruned, at least to the extent of removing dead branches. And because they are taken care of, they tend to produce the best piñon nuts and those easiest to harvest. Ultimately, a *dehesa* can best be understood as pastureland with a mosaic of uses, including grazing and, when necessary, dry farming.

EJIDO Outskirts of a village, from the Latin *exitus,* the way out. This land, which is neither planted nor worked, serves as the commons and contains four main divisions that at times interweave like a braid: *sierras*, *montes*, *dehesas*, and *solares*. See *dehesas*, *montes*, *sierras*, and *solares*.

ERA Particular land formations, from the Latin *aera* or possibly the Hebrew *erez,* both of which mean land. There are two types of *eras*—one for threshing grains and the other for use as a seed plot, traditionally in the form of a sunken bed. Among the Zuni Indians, such beds are known as "waffle gardens" and usually contain recent plantings of lettuce, radishes, and other vegetables; elsewhere they are known as Afghan gardens and resemble a comb. Both types of beds are still found in New Mexico and also along the outskirts of Chihuahua City, Mexico.

JOYA Happiness, from the Tuscan word *gioia,* which also means something very precious, such as a jewel. In agriculture, this term came to signify the most fertile land, the place where people would plant their chile and other vegetables for home use or to trade for crops they did not grow. Intimately familiar with the microclimate of their land, farmers knew exactly where to plant each crop for best results.

LATIFUNDIO A rustic country estate with extensive land holdings, from the Latin *latifundium.* Often these estates were used for livestock grazing.

MAYORDOMO From the Latin *maior,* meaning elder, and *domus,* of the house. In an agricultural context, it means the one in charge of the *acequia madre.* Not much can be found about such an individual in Moorish sources. Details begin appearing in later Christian documents, an example of which, from the thirteenth century, portrays the *mayordomo* as one who takes care of the water. Under Spanish influence, the *mayordomo* came to be known as a *cequiero,* one who divides the water, and acted as a bartender making sure everyone had been served. Over time, this individual became associated with the law and regarded as a *juez de agua,* or water judge—in effect, one in charge of delivering the water. Elected along with the *comisión,* he operated under its direction. In New Mexico, the connotation is more spiritual. The *mayordomo* is referred to as one who is *"digno de confianza"* ("worthy of being trusted"), *"el que es fiel"* ("he who is faithful"), and *"el fiel del agua"* ("faithful with the water").

MELGA (MIELGA) Originally alfalfa, a common pasture plant for animals, from a corruption of the Latin *medica herba.* In the new world, the term came to denote a parcel of land where alfalfa was planted. Reduced to manageable strips, these parcels each measured about fifty feet long with a width of a *suerte.* Today, a *melga* that is part of a *joya* can be further divided into *eras,* as a water conservation strategy. A *melga* is also a tool for measuring water, used commonly in the Alpujarras, south of Granada, Spain. See also *alfalfa.*

MERCED A favor or gift usually presented by a king to his subjects, from the Latin *merces.* In the Americas, it referred to a certain amount of land granted by the king of Spain in an effort to populate the northern frontier of New Spain. In New Mexico, about three hundred *mercedes de tierra,* or land grants, were made between 1598 and 1846 under Spanish and Mexican rule. A smaller number of *mercedes de agua,* or water grants, were also made, establishing a free distribution of water for irrigation.

MILPA A place where corn is planted, from the Náhuatl *milli,* meaning seedbed, and *pa,* a place where corn is sown. This was strictly a Mesoamerican idea. In fact, corn was introduced into Spain only after the colonizers' encounter with Mesoamerica, and then considerable time passed before it was considered anything more than feed for animals. Even today, residents of Spain do not eat corn on the cob.

MINIFUNDIO A very small piece of privately owned cultivated land resulting from a breakdown in the feudal system of rural landscape ownership that occurred under Arabic rule, from the Latin *mini,* meaning very small, and *fundio*, legal property. Under the Moors, a *minifundio*, which usually measured about a hectare (two and a half acres), could feed a family of four, but following the conquest of the Castilians, who were unfamiliar with this type of irrigated farming, it could barely feed one person. See also *latifundio.*

MONTE Mountain, from the Latin *mons.* It is said that mountains are pregnant, because of their swellings and bulges, and because they have a head (the summit), shoulders (the watershed slopes), and *faldas,* or skirts. Mountains also have *cejas,* (eyebrows), *chuchillas* (knife edges), and *cordilleras* (smaller mountain ranges extending from them).

NORIA A hole in the ground, from which water is drawn, from the Arabic *na'ura,* meaning well. At one time, water was hauled up by a *cigüeña* (crank or wench), while now it is done by electric motor.

PARCIANTE A small piece of land under a private ownership, from the French *parcelle*, which is derived from the Latin *particella.* In New Mexico, the term refers to a property owner who has water rights. In other Spanish-speaking areas, a *parciante* is known as a *parcelario,* a person who owns a piece of land.

PEON An acequia worker or a measure of irrigation time, from the Latin *paeon,* meaning day laborer. The first description alludes to a laborer engaged in the acequia's annual spring cleanup or in cutting willows along its banks. The second description pertains to water rights; in the old world, it generally took two workers five days apiece to irrigate a *suerte,* while in the new world a *peon* represents a full day of work, from 8:00 a.m. to 5:00 p.m. The amount of land owned by a *parciante* determines the portion of a *peon* needed to irrigate it.

PÍCARO Since there was little if any work to be had in early-sixteenth-century Spain, many young men, especially those living along the Camino de Santiago, took temporary jobs in the *mesones,* or inns, *picando,* slicing and dicing vegetables, and working with the clergy as well, giving rise to a group of very astute, street-smart young men called "*picadores.*" Over time, the *picadores* evolved into *pícaros,* today's beloved folk heroes of Spanish literature, including such luminaries as Lazarillo de Tormes and Pedro de Urdemalas, who made his way first to Chile and on to New Mexico.

REPARTIMIENTO DE AGUA Sharing of water, based on the amount of water in the river and the number of acres under cultivation—a notion rooted in the Moorish concept of *equidad*, from the Qur'an. When a *repartimiento* goes into effect, the *comisionados* and *mayordomos* first calculate the *surcos* of water currently in the river.

They then divide this figure by the number of participating acequias (the Río Embudo, for instance, has eight acequias), whereupon the water is distributed equally among the acres requiring irrigation.

SANGRADA (SANGRIA) A small ditch used to drain a *ciénaga,* or marshland, so it can be cultivated, originating in folk wisdom concerning the human body. In much the same way that a person *sangrars* (drains) a painful injury, so does an agriculturalist treat a parcel of land with too much water—by draining it via a *sangría*. In contemporary usage, a *sangría* is also a small ditch that branches laterally for purposes of watering a *melga*.

SIERRAS A mountainous terrain with features resembling the teeth of a saw, derived from the Latin *serra*. From Spain, where the term applies to such mountains, it was transported through Mesoamerica and used to name similarly shaped mountain ranges in the new world. *Sierras* are often seen as "keepers of the water," since slowly melting snow from their peaks eventually forms watersheds, not only providing water for irrigation but also recharging the aquifers.

SOLARES Sites, measuring approximately one hundred forty feet squared, where individual houses were often built, from the Spanish word *suelo,* meaning to construct the floor of a house; also, spaces between an acequia and the commons, where settlers built their houses, *corrals, gallineros* (chicken coops), and *trochiles* (pigpens), and kept their *leña* (woodpile). As for the houses, they were usually either L-shaped or U-shaped, as were Moorish houses in the *alquerías*. Most house complexes also included a *dispensa,* or utility room, and a *soterrano*, root cellar, where foods were stored during the winter.

SUERTE A long-lot of approximately thirteen acres granted to a *vecino,* or settler, by lottery or luck (*suerte*), a practice originating in the Middle East. The private land was below an acequia, which both irrigated it and divided it from the commons. This type of land distribution ensured that everyone had good soil close to home for growing crops and raising domestic animals, most often a milk cow and a few sheep.

Initially, a settler was granted up to three *suertes,* and if he maintained them well over three years, he could apply for three more as part of the grant. Four years later, he could then buy as many as he could afford. In some places, such as the San Luis Valley of southern Colorado and elsewhere throughout the Río Arriba bioregion, the expanded *suertes* continued above the acequia. In such instances, the *suerte*s came to be known as *extensiones* (extensions) or as *tiras* (strips).

SURCO A unit of measure used to calculate the amount of water in a river. In the Southwestern United States, a *surco de agua* is the amount of water that can flow through a *buje*, the approximately five-inch opening in the center of a cartwheel. A *buey* (ox) *de agua* is the amount of water that can flow under an ox standing at the point in the river where the water is diverted; there are forty-eight *surcos* in a *buey de agua.* A *surco* also refers to the amount of water needed to irrigate a *suerte* of land in twenty-four hours.

TAPANCO A pile of dirt, from the Náhuatl *tlapantli*. This heap of soil is used to divert water from a secondary acequia to the furrows that need to be irrigated on a *suerte* of land.

TERRAZAS Terraces irrigated by diverting water from an acequia, derived from the Latin *terraceus*. Terraces on slopes are called *bancales*; those in valleys are known as *bancos*, and those alongside a meandering river are referred to as *ancones*. Wherever they are located, they provide for effective water use, prevent erosion and thus build up fertility of the soil, and make marginal lands available for farming.

VEGA A word for land, variously defined. It could be derived from the Iberian *vaica*, meaning fertile plain, or from the Latin *vigore*, vigorous and fertile. It is also said to be an Arabic word that describes a piece of land used for growing crops. In Spain and most Latin American countries, it refers to a lowland area that is level and fertile. In Chile, it refers to a moist piece of land. In Andalucía, it denotes a place where food is grown, such as the *vegas* of Granada, Valencia, and Murcia. In New Mexico, it refers more specifically to an irrigated pasture.

VEREDA A narrow trail used by horses to move herds or smaller flocks of livestock for purposes of grazing, from the Latin *vereda,* meaning narrow road. Such trails can zigzag wildly, which accounts for the *dicho "Quien deja el camino real por la vereda, piensa atajar y rodea."* ("He who leaves the royal road for the trail thinks he is taking a shortcut but instead makes the trip longer.")

INDEX

R

S